Plants as a Source of Natural Antioxidants

Plants as a Source of Natural Antioxidants

Editor

Surendra Mohite

scitus
academics

Plants as a Source of Natural Antioxidants

Edited by **Surendra Mohite**

Printed in 2017

ISBN: 978-1-68117-088-6

Library of Congress Control Number: 2015935487

Notice

Reasonable efforts have been made to publish reliable data and views articulated in the chapters are those of the individual contributors, and not necessarily those of the editors or publishers. Editors or publishers are not responsible for the accuracy of the information in the published chapters or consequences of their use. The publisher believes no responsibility for any damage or grievance to the persons or property arising out of the use of any materials, instructions, methods or thoughts in the book. The editors and the publisher have attempted to trace the copyright holders of all material reproduced in this publication and apologize to copyright holders if permission has not been obtained. If any copyright holder has not been acknowledged, please write to us so we may rectify.

Contents

vi

Preface

Plants have a significant role in maintaining human health and improving the quality of human life for thousands of years and providing valuables in the form of medicines, beverages, cosmetics and dyes. There exists a plethora of knowledge and information and benefits of herbal drugs in our ancient literature of Ayurvedic and Unani medicine. Diseases that remain most challenging in today's healthcare system tend to be complex involving multiple mechanisms, targets and drugs for effective disease management. In contrast to current combination therapies, however, plant based drugs contain a mixture of multiple components thereby saving considerable time and expense. Numerous and diverse classes of natural products have been isolated and their structures characterized in the past century. The elucidation of biological and biochemical mechanisms of natural products with therapeutic action have been invaluable to the efforts of organic and medicinal chemists as tools for deciphering the logic of biosynthesis and as platforms for developing frontline drugs. Plants have limitless ability to synthesize aromatic secondary metabolites, important subclasses in this group of compounds include phenols, phenolic acids, quinones, flavones, flavonoids, flavonols, tannins and coumarins. These groups of compounds show antimicrobial effect and serves as plant defense mechanisms against pathogenic microorganisms. The importance of natural antioxidants has been clarified by numerous studies which have demonstrated that the consumption of foods rich in such phytochemicals can exert beneficial effects upon human health, possibly by interfering in the processes involved in reactive oxygen and nitrogen species mediated pathologies. Harmful side effects and weak effectiveness of curative agents in market has made their use limited and the search to find more effective agents continues. Investigation in the plant kingdom culminated in the discovery of many herbal hypoglycemic agents.

Editor

Correlation of Antiangiogenic, Antioxidant and Cytotoxic Activities of some Sudanese Medicinal Plants with Phenolic and Flavonoid Contents

Loiy Elsir A Hassan[1, 2,] Mohamed B Khadeer Ahamed1, Aman S Abdul Majid[3,] Hussein M Baharetha[1, 4,] Nahdzatul S Muslim[1,] Zeyad D Nassar[5,] and Amin MS Abdul Majid[1]

[1]EMAN Research and Testing Laboratory, School of Pharmaceutical Sciences, Universiti Sains Malaysia, Penang, Malaysia

[2]Department of Botany, Faculty of Science & Technology, Omdurman Islamic University, Omdurman, Sudan

[3]Advanced Medical and Dental Institute (IPPT), Universiti Sains Malaysia, Penang, Malaysia

[4]Department of Pharmacy, College of medicine and Health Sciences, Hadhramout University, Fuluk, Mukalla, Hadhramout, Republic of Yemen

[5]School of Pharmacy, the University of Queensland, 20 Cornwall Street, Woolloongabba, QLD 4102, Australia

ABSTRACT

Background

Consumption of medicinal plants to overcome diseases is traditionally belongs to the characteristics of most cultures on this earth. Sudan has been a host and cradle to various ancient civilizations and developed a vast knowledge on traditional medicinal plants. The present study was undertaken to evaluate the antioxidant, antiangiogenic and cytotoxic activities of six Sudanese medicinal plants which have been traditionally used to treat neoplasia. Further the biological activities were correlated with phytochemical contents of the plant extracts.

Methods

Different parts of the plants were subjected to sequential extraction method. Cytotoxicity of the extracts was determined by dimethylthiazol-2-yl) - 2, 5 diphenyl tetrazolium bromide (MTT) assay on 2 human cancer (colon and breast) and normal (endothelial and colon fibroblast) cells. Anti-angiogenic potential was tested using *ex vivo* rat aortic ring assay. DPPH (1, 1-diphenyl-2-picrylhydrazyl) assay was conducted to screen the antioxidant capabilities of the extracts. Finally, total phenolic and flavonoid contents were estimated in the extracts using colorimetric assays.

Results

The results indicated that out of 6 plants tested, 4 plants (*Nicotiana glauca, Tephrosia apollinea,Combretum hartmannianum* and *Tamarix nilotica*) exhibited remarkable anti-angiogenic activity by inhibiting the sprouting of microvessels more than 60%. However, the most potent antiangiogenic effect was recorded by ethanol extract of *T. apollinea* (94.62%). In addition, the plants exhibited significant antiproliferative effects against human breast (MCF-7) and colon (HCT 116) cancer cells while being non-cytotoxic to the tested normal cells. The IC_{50} values determined for *C. hartmannianum*, *N. gluaca* and *T. apollinea* against MCF-7 cells were 8.48, 10.78 and 29.36 µg/ml, respectively. Whereas, the IC_{50} values estimated for *N. gluaca*, *T. apollinea* and *C. hartmannianum* against HCT 116 cells were 5.4, 20.2 and 27.2 µg/ml, respectively. These results were more or less equal to the standard reference drugs, tamoxifen (IC_{50} = 6.67 µg/ml) and 5-fluorouracil (IC_{50} = 3.9 µg/ml) tested against MCF-7 and HCT 116, respectively. Extracts of *C. hartmannianum* bark and *N. glauca* leaves demonstrated potent antioxidant effect with IC_{50s} range from 9.4–22.4 and 13.4–30 µg/ml, respectively. Extracts of *N. glauca* leaves and *T apollinea*aerial parts demonstrated high amount of flavonoids range from 57.6–88.1 and 10.7–78 mg quercetin equivalent/g, respectively.

Conclusions

These results are in good agreement with the ethnobotanical uses of the plants (*N. glauca*, *T. apollinea*, *C. hartmannianum* and *T. nilotica*) to cure the oxidative stress and paraneoplastic symptoms caused by the cancer. These findings endorse further investigations on these plants to determine the active principles and their mode of action.

BACKGROUND

Cancer is a major public health burden in both developed and developing countries. The International Agency for Research on Cancer (IARC), the specialized cancer agency of the World Health Organization reported that about 14.9 million cancer cases were estimated around the world in 2013, of these 7.7 million cases were in men and 6.9 million in women and further this number is expected to increase to 24 million by 2035 [1]. Treating cancer has become a major challenge as there is no single effective treatment that works for all types of cancer. Most of conventional chemotherapy regimens which employ different combinations of cytotoxic drugs which are often associated with serious side effects and chemoresistance. Conventional therapy has also become less favorable in the mindset of sufferers and as a result many patients resort to seeking alternative treatments [2]. The resistance of metastatic cancerous cells to chemotherapy and its adverse effects has become a serious challenge in cancer research. Despite the intensive progress in chemotherapeutics in the last decades, the need to discover and to develop new, alternative, or adjuvant therapeutic agents remains.

Botanicals have long been used traditionally in treatment of various types of cancers [3] and often less associated with the side effects like the modern chemotherapy has [4]. Realizing the potential benefits of botanicals as a source of active anticancer compounds, the National Cancer Institute (U.S.A) collected about 35,000 plant samples from 20 countries and has screened around 114,000 extracts for anticancer activity [5]. Out of the 92 anticancer drugs marketed prior to 1983 in US and among the ones sold worldwide between 1983 and 1994, 60% are of natural origin [6]. This includes natural products, derivatives of natural products or semi-synthetic pharmaceuticals based on natural products models [7].

Angiogenesis is the sprouting of new blood vessels from pre-existing vessels and is strongly implicated in solid tumorgenesis, proliferative retinopathies, obesity and rheumatoid arthritis [8]. Tumor angiogenesis is the consequence of an angiogenic imbalance in which proangiogenic factors predominate over antiangiogenic factors. Furthermore, angiogenesis is essential for growth and metastasis of malignant tumors. Vascular Endothelial Growth Factor-A (VEGF-A) is believed to be a critical angiogenic mitogen [9]. Therefore, tumor angiogenesis can be considered as an important pharmacological target for cancer prevention and treatment [10, 11]. Consequently, this hypothesis has paved a pathway for the development of the cutting edge therapeutic technology called angiotherapy. Anti-angiogenic approach can overcome the cytotoxic adverse effects and chemoresistant problems associated with the classical chemotherapies. Anti-angiognenic drugs work by inhibiting the synthesis of new blood vessels that supply blood, nutrients and oxygen to growing tumor. Previous reports of Avastin, a monoclonal antibody for VEGF, and fluorouracil-based combination therapy showed a significant improvement in survival of patients with metastatic colorectal carcinoma [12]. However conventional antiangiogenic compounds based on monoclonal antibody technology may have limitations from the cost point of view. Plant sources of antiangiogenic compounds have been explored as they are more economical to produce in large scale [13]. However at present there are no plant based antiangiogenic compounds available commercially in the market.

Plants have many phytochemicals which are potential source of natural antioxidants, such as phenolic diterpenes, flavonoids, tannins and polyphenolic acids [14] with versatile biological activities. Plant polyphenolics have been recognized as the potential therapeutic agents targeting cancer, pathological angiogenesis and cardiovascular disease in the next decade [15, 16]. These benefits have been attributed to the presence of antioxidant-rich polyphenolic compounds [16, 17].

The use of traditional medicine especially medicinal plants in Sudan is still the main alternative therapy which is based entirely on the indigenous knowledge gained from ancestral experience. Although, there is an important local ethno botanical biography describing the most frequently used plants in treatment of various clinical conditions however, very few have been studied scientifically for chemometric analysis of the medicinal herbs to identify the active principles. The vast majority is still unexplored phytochemically and their medicinal properties have not yet validated [18].

Compelling data implicate angiogenesis and tumor-associated neovascularization as a central pathogenic step in the process of tumor growth, invasion and metastasis. Subsequently, it was shown that a significant correlation existed between the degree of tumor angiogenesis (micro vessel density) and survival in patients presenting with lymph node-negative breast carcinoma [19]. Therefore, it comes to handy that targeting tumor angiogenesis using antiangiogenic agents the blood vessels which supply the tumor with nutrients and oxygen, in turn it could lead to halt tumor growth and metastasis.

With this background, the present study was undertaken to analyze the antiangiogenic, antioxidant and cytotoxic properties of 32 extracts prepared from six Sudanese plants; *Indigofera spinosa*Forsk. (Leguminosae), *Nicotiana glauca* var. (Solanaceae), *Tephrosia apollinea* (Del.) Link (Leguminosae), *Tamarix nilotica* (Ehrenb.) Bunge (Tamaricaceae), *Combretum hartmannianum*Schweinf. (Combretaceae) and *Capparis decidua* (Forsk) Edgew (Capparaceae). This study is the first to report the antiangiogenic properties of these selected Sudanese medicinal plants and correlated the activity with antioxidant property. In addition, an investigation on the cytotoxicity of the extracts was conducted to identify the potential source of antineoplastic agents.

METHODS

Plant Material

Six Sudanese medicinal plants, *I. spinosa*, *N. glauca*, *T. apollinea*, *T. nilotica*, *C. hartmannianum* and *C. decidua* were selected for the study. Plant material was collected during the period of March-July 2013 except *C. hartmaniunum* which was collected during March 2014 from Elgadarif City – Sudan. The taxonomic authentication of all the plants was carried out at The Medicinal and Aromatic Plants Research Institute, National Center for Research by Dr. Wail Alsadig. Voucher specimens (voucher references numbers: MAPRI/NB-53a-g) were deposited at the herbarium of the institute.

Preparation of Extracts

The plant materials were dried in oven (35–40°C) and powdered mechanically. The pulverized plant material (50 g) was subjected to sequential extraction method started with n-hexane and followed by ethanol, methanol and water. All the extracts were prepared by 250 ml of the solvents using hot maceration (40°C) method with intermittent shaking. The extracts were filtered and concentrated at 45°C under vacuum by rotary evaporator (Buchi, USA) and further dried overnight at 45°C. Stock solutions of the extracts were prepared at 10 mg/ml in 100% dimethyl sulfoxide (DMSO). Further serial dilution of the stock was performed with cell culture media to obtain a range of desired concentrations of the extracts. All solvents used in this study were of analytical grade.

Experimental Animals

Twelve to fourteen weeks old healthy Sprague Dawley male rats were used. To avoid physiological variations that could affect the process of angiogenesis in female rats due to estrous cycle, only male rats were used in rat aortic ring assay. The animals obtained from animal house facility of Universiti Sains Malaysia (USM) and were kept for one week in animal transit house (School of Pharmaceutical Sciences, USM) prior to the experiments. The animals were kept in well ventilated cage with food and water provided. The animals were euthanized using CO_2 and dissected to excise thoracic aorta. All procedures were carried out according to the guidelines of Animal Ethics Committee USM. The present study was submitted to the institutional animal ethics committee, "Animal Ethics Committee USM" for evaluation and the present study is approved by the committee (approval Reference number: PPSG/07 (A)/044/ (2010) (61)).

Chemicals and Reagents

Cell culture reagents were purchased from Gibco, USA; RPMI 1640 medium; catalogue number (A10491-01), Dulbecco's Modified Eagle Medium; Catalogue number (31100–035) were obtained from GIBCO, UK. Phosphate buffered saline, trypsin, heat inactivated foetal bovine serum (HIFBS), penicillin/streptomycin (PS), fibrinogen, aprotinin, thrombin, suramin, aprotinin, 6-Aminocaproic acid, L-glutamine, thrombin and gentamicin were purchased from Sigma, Germany. MTT (3-(4, 5-Dimethylthiazol-2-yl) - 2,5diphenyl tetrazolium bromide) was procured from Sigma-Aldrich, USA. Dimethyl sulfoxide (DMSO) was purchased from Fluka, USA.

Cell Lines and Culture Conditions

Human Umbilical Vein Endothelial Cell line HUVEC (Passage No. 3), catalogue number (C2517A); human colorectal carcinoma cell line HCT-116 (Passage No. 5), catalogue number (CCL-247); human hormone sensitive and invasive breast cancer cell line MCF-7 (Passage No. 4), catalogue number (HTB-22); human colorectal normal cell line CCD-18 (Passage No. 3), catalogue (CRL-1459) were purchased from

ScienCell, USA. HUVEC were maintained in endothelial cell medium (ECM) (ScienCell, USA) supplemented with endothelial cell growth supplements (ECGS), 5% HIFBS and 1% PS. HCT-116 cells were maintained in RPMI whereas, MCF-7 and CCD-18Co were maintained in DMEM medium. The media were supplemented with 5% heat inactivated fetal bovine serum and 1% penicillin/streptomycin. Cells were cultured in a humidified incubator at 37°C supplied by 5% CO_2. Cell culture work was done in sterile conditions using Class II biosafety cabinet (ESCO, USA).

Rat Aorta Ring Assay

This assay was carried out on rat aortic explants as previously described [20]. Thoracic aortas were removed from euthanized male rats, rinsed with serum free medium and cleaned from fibroadipose tissues. Totally 18 rats were used in this assay and approximately 12 to 14 rings (each ring is about 1 mm thickness) were prepared from an each aorta. The aortas were cross sectioned into small rings and seeded individually in 48-wells plate in 300 µL serum free M199 media containing 3 mg/ml fibrinogen and 5 mg/ml aprotinin. Ten microliters of thrombin (50 NIH U/ml in 1% bovine serum albumin in 0.15 M NaCl) was added into each well and incubated at 37°C for 90 min to solidify. A second layer (M 199 medium supplemented with 20% HIFBS, 0.1% έ-aminocaproic acid, 1% L-Glutamine, 2.5 µg/ml amphotericin B, and 60 µg/ml gentamicin) was added into each well (300 µL/well). All the extracts were added at final concentrations of 100 µg/ml. Suramin and 1% DMSO were used as positive and negative controls, respectively. On day four, the medium was replaced with a fresh one containing the test materials. On day five, aortic rings were photographed using EVOS f1 digital microscope (Advanced Microscopy Group, USA) (40× magnification) and subsequently the length of blood vessels outgrowth from the primary tissue explants was measured using Leica Quin software.

The inhibition of blood vessels formation was calculated using the formula;

% blood vessels inhibition = $[1 - (A0/A)] \times 100$, Where; A0 = distance of blood vessels growth in treated rings in µm, A = distance of blood vessels growth in the control in µm.

The results are presented as mean percent inhibition ± SEM, (n =8).

The significant difference between the microvessels out growth in treated versus untreated aortic rings was calculated using Student's t test. Based on the results of this assay, TAF273 was chosen for the subsequent investigations for the anti-angiogenic property.

Cytotoxicity Assay

The MTT cytotoxicity assay was performed according to the method previously described [21]. Cells were seeded at 1.5×10^4 cells in each well of 96-well plate in 100 µl of fresh culture medium and were allowed to attach for overnight. For screening, the cells (70 - 80% confluency) were treated with the extracts at the final concentration of 50 µg/ml. Later on, in order to obtain a dose–response curve, the most active extracts were tested for cytotoxicity at 3.12, 6.25, 12.5, 25, 50 and 100 µg/ml concentrations. After 48 h of the treatment the medium was aspirated and the cells were exposed to MTT solution prepared at 5 mg/ml in sterile PBS was added to each well at 10% v/v in the respective medium and was incubated at 37°C in 5% CO_2 for 3 h. The water insoluble formazan salt was solubilized with 200 µl DSMO/well. Absorbance was measured by infinite® Pro200 TECAN Group Ltd., (Switzerland) at primary wave length of 570 nm and reference wavelength of 620 nm. Each plate contained the samples, negative control and blank. DMSO (1% v/v) was used as a negative control. 5-fluorouracil, Tamoxifen and Betulinic acid were used as standard reference control for HCT 116, MCF-7 and CCD-18Co cell lines, respectively. The assay was performed in quadricate and the results were presented as a mean percent inhibition to the negative control ± SEM.

Determination of Total Phenols

Total phenols in the extracts were determined by a colorimetric method as described by Al-Suede and co-workers [22]. A stock of 1 mg/ml of extracts was prepared in methanol and 100 µl of each extract was added separately to 750 µl of Folin-Ciocalteau phenol reagent (1:10 diluted with double distilled H_2O). After 5 min incubation in the dark at room temperature, 750 µl sodium bicarbonate solution (60 g/l) was added and incubated at 30°C in the dark for 90 min. The absorbance was measured at 725 nm using TECAN Multi-mode microplate reader Model Infinite® 200 (Mannedorf, Switzerland). Gallic acid was used (5–80 µg/ml) to construct the standard calibration curve. The results were expressed as Gallic acid equivalents per 100 mg of extract (mg GAE/100 mg).

Determination of Total Flavonoids

The total flavonoids content in the extracts was determined using aluminum chloride colorimetric method with quercetin as standard [23]. A solution of 4 mg/ml of quercetin in methanol was prepared. Exactly, 500 µl of different concentrations (3 to 200 µg/ml) of the extracts were taken in separate test tubes. To each of the test tubes, 0.1 ml of 10% (w/v) aluminum chloride solution, 0.1 ml of 1 M potassium acetate solution, 1.5 ml of methanol and 2.8 ml of distilled water was added. The test tubes were thoroughly mixed and after incubation at room temperature for 30 min, the absorbance reading of the reaction mixture was measured at 415 nm using a spectrophotometer (Perkin Lambda 45). A standard curve plotting all the different concentrations of quercetin standard was constructed and the total flavonoid content is expressed as micrograms of quercetin equivalent. The data were presented as mean ± SEM (n =6).

DPPH Scavenging Effect

DPPH (1, 1-diphenyl-2-picrylhydrazyl) assay was carried out to evaluate the scavenging activity of the extracts [24]. The stock solution of DPPH was prepared at a concentration of 200 µM in absolute methanol while stock solutions of the extracts were prepared at concentration of 10 mg/ml. DPPH was dispensed into 96-well plate (100 µl/well) and immediately, 100 µl of test samples were added at final concentrations of 12.5, 25, 50, 100, 200 µg/ml. Methanol alone and methanol with DPPH were used as blank and negative control, respectively. Ascorbic acid was used as positive control. The mixtures were incubated at 30°C for 30 min in the dark and then the absorbance was measured at 517 nm using TECAN microplate reader Model Infinite® 200 (Mannedorf, Switzerland). The dose response curves were obtained and then used to calculate the median inhibitory concentration (IC_{50}). The results are expressed as mean $\pm$ SEM (n =6).

RESULTS

Plant Extraction

Four exacts were prepared from each plant material, starting with n-hexane followed by ethanol, methanol and water. The yield of each extract was calculated and presented in Table 1 as w/w percent yield. Among all the extracts, hexane extracts of all the tested plants produced the lowest yield except for *N. gluaca* leaves extract (7.8%). On average, the ethanol extracts of the tested plants showed the highest yield followed by methanol and water extracts. For instance, the highest yield recorded was 15.68% for ethanol extract of *C. hartmannianum* leaves. Table 1 shows the list of plants and the parts used in this study.

Table 1: Parameters of different extracts of the selected Sudanese plants

Botanical name	Local name	Used part	Solvent	Texture	Yield %
Indgosfera spinosa	Hipatoheep	Stem	n-Hexane	Sticky	0.8
			Ethanol	Solid	2.82
			Menthol	Solid	1.72
			Water	Powder	3.36
Nicotiana gluaca	Hili leasa	Leaves	n-Hexane	Sticky	7.8
			Ethanol	Gummy	11.73
			Menthol	Gummy	4.11
			Water	Powder	1.698
		Stem	n-Hexane	Gummy	0.82
			Ethanol	Gummy	6.21
			menthol	Gummy	3.94
			water	Powder	5.94
Tephrosia apollinea	Dhawasi	Aerial parts	n-Hexane	Gummy	0.6
			Ethanol	Sticky	2.41
			Menthol	Sticky	5.86
			Water	Powder	4.01

Tamarix nilotica	Tarffa	Leaves	n-Hexane	Gummy	1.4
			Ethanol	Gummy	12.64
			Menthol	Gummy	4.14
			Water	Powder	7.41
Combretum hartmannianum	El-Habeel	Leaves	n-Hexane	Gummy	7.16
			Ethanol	Sticky	15.68
			Menthol	Gummy	6.64
			Water	Powder	6.04
		Bark	n-Hexane	Gummy	0.163
			Ethanol	Solid	9.18
			Menthol	Solid	3.37
			Water	Powder	2.66
Capparis decidua	Tunduli	Stem	n-Hexane	Gummy	0.892
			Ethanol	Gummy	6.81
			Menthol		3.37
			Water	Powder	01.56

Hassan *et al.*

Hassan *et al.* *BMC Complementary and Alternative Medicine* 2014 14:406, doi:10.1186/1472-6882-14-406

Inhibitory Effect of the Extracts on Sprouting of Microvessels in Rat Aortic Explants

This assay was performed as the primary assay to screen the antiangiogenic potential of the extracts. Figure 1A shows a massive sprouting of microvessels in the aortic explants of vehicle treated group (negative control). Plant extracts with more than 60% inhibition of sprouting of blood vessels were considered as active extracts. Table 2 depicts the antiangiogenic properties of plant extracts determined by rat aorta ring assay. Out of 32 extracts, 12 extracts from three plants namely *T. apollinea*, *C. hartmannianum* and *T. nilotica* showed potent (more than 60%) inhibitory activity. The highest inhibition (100%) was produced by ethanol extract of aerial parts of *T. apollinea* (Figure 1B), followed by water extracts of *T. nilotica* (Figure 1C).

Interestingly, all the extracts of *C. hartmannianum* (Figure 1D) stem bark displayed significant (*p* <0.05) inhibition of micro-blood vessels with an order of ethanol (88.74%) followed by menthol (86.68%) and n-hexane (81.75%). Moreover, methanol and ethanol extracts of *C. hartmannianum* leaves also demonstrated significant (*p* <0.01 and *p* <0.05, respectively) antiangiogenic effect with 68.84 and 63.68%, respectively. In addition, Figure 1E depicted the remarkable inhibitory effect on microvessel growth from the rat aortic explant treated with *T. nilotica* extract. These results were very much comparable with the positive control, suramin which demonstrated potent inhibition of microvessel growth (Figure 1F). Figure 1G graphically depicts the difference between the effects of the extracts and Suramin.

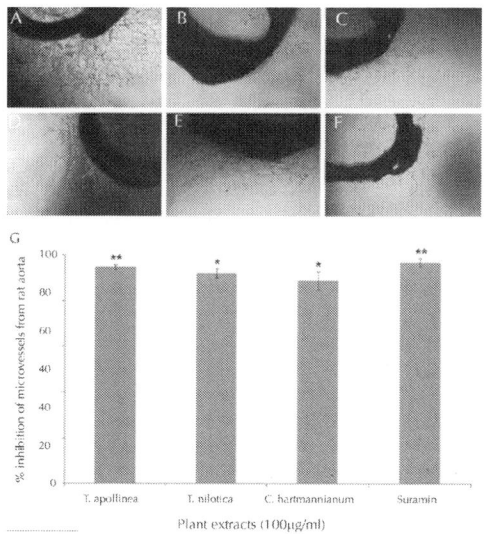

Figure 1: Antiangiogenic effect of selected Sudanese plants (100 µg/ml) against sprouting of microvessels in rat aortic explants. A) Photomicrographic image of rat aortic ring of negative control showing extensive growth of microvessels. B) Photomicrographic image of rat aortic ring representative of polar extract of *T. apollinea* displaying potent inhibition of growth of microvessels. C) Treatment with *T. nilotica* showed significant inhibitory effect against sprouting of microvessels from rat aortic rings. D) Photomicrographic image of rat aortic explant treated with *C. hartmannianum* extract demonstrated considerable effect of antiangiogenicity. E) Photomicrographic image of rat aortic explant treated with non-polar extract of *T. apollinea* showing significant inhibitory effect on growth of microvessels. F) Photomicrographic im-

age of rat aortic explant treated with standard reference, suramin exhibiting strong inhibitory effect on microvessels growth. G) Graphical representation of antiangiogenic activity of the active extracts obtained from the selected Sudanese plants. The values are expressed as mean ± SEM (n =10). **$p < 0.01$ and *$p < 0.05$ compared to negative control group (0.1% DMSO).

Table 2: Antiangiogenic activity of the selected Sudanese plants on rat aortic explants

Botanic namen	Part used	Extract	Inhibition (%)
Tephrosia apollinea	Aerial parts	Ethanol	94.6 ± 1.8**
Tamarix nilotica	Leaves	Water	91.7 ± 2.2**
Combretum hartmannianum	Bark	Ethanol	88.7 ± 1.6*
Combretum hartmannianum	Bark	Methanol	86.6 ± 3.8**
Combretum hartmannianum	Leaves	n-hexane	81.7 ± 1.7*
Tephrosia apollinea	Aerial part	Methanol	79.8 ± 2.6*
Tephrosia apollinea	Aerial parts	n-hexane	73.1 ± 1.5*
Combretum hartmannianum	Bark	Water	71.5 ± 5.3*
Tamarix nilotica	Leaves	Methanol	65.5 ± 2.9*
Combretum hartmannianum	Leaves	Methanol	68.8 ± 2.2*
Combretum hartmannianum	Leaves	Ethanol	63.6 ± 1.4*

Results are presented as mean percent inhibition ± S.D, n = 3.

Statistical significance is expressed as * = $P < 0.05$, ** = $P < 0.01$.

Hassan et al.

Hassan et al. BMC Complementary and Alternative Medicine 2014 14:406, doi:10.1186/1472-6882-14-406

Anti-Proliferative Effect of the Extracts against Cancer Cells

The MTT assay was used to screen the possible cytotoxic activity of 32 extracts against two human cancer cells lines (HCT-116 and MCF-7) and two normal cell lines (HUVEC and CCD-18Co). For screening, the cells were treated with the extracts at 50 μg/ml concentration. The extracts with more than 60% inhibition of cell proliferation were considered as active extracts. Hexane extracts of *N. glauca* leaves and stem exhibited the highest cytotoxicity on all the tested cell lines, while hexane and ethanol extracts of aerial part of *T. apollinea* showed selective antiproliferative effect against breast cancer cell line (MCF-7) with 87.20 and 86.94%, respectively. The hexane extract ofC. *hartmannianum* leaves potently inhibited the growth of both cancer cell lines (HCT-116 and MCF-7) with 80.03 and 95.76% anti-proliferative effect, respectively. Moreover, the ethanol extract of *C. hartmannianum* stem bark showed selective cytotoxicity towards MCF-7 with 76.79%. Interestingly, all the extracts of *C. hartmannianum* showed poor cytotoxicity against the normal cell lines (Table 3). Further, the most active extracts were selected to study the dose response cytotoxic effect. The median inhibitory concentration (IC_{50}) values for the most active extracts and the respective standard reference drugs were calculated for all the tested cell lines and the values are given in Table 4.The results were comparable with the respective standard reference drugs, 5-fluorouracil, betulinic acid and tamoxifen. Figures 2A and B show the graphical illustration of the dose-dependent antiproliferative effect of the active extracts against human HCT 116 and MCF-7 cell lines.Figure 3 shows the photomicrographic images of the treated HCT 116, MCF-7, CCD-18Co and HUVEC cell lines. The morphological feature of the treated cancer cells presented clear evidence of strong cytotoxicity of the extracts, as the vehicle (1% DMSO) treated cells displayed a compact monolayer of aggressively growing cancer cells with prominent nuclei and intact cell membrane. Whereas the images taken from the extracts treated group showed a drastic reduction in the number of cells because of the anti-proliferative activity of the extracts. In addition, the extracts severely affected the pseudopodial projections of the cells which rendered the cells non-adherent and become round shaped. Interestingly, all the extracts studied showed either mild or negligible cytotoxicity towards the both normal (HUVEC and CCD-18Co) cell lines which were used as the model cell lines for the normal human cells.

Table 3: Cytotoxic effect of different extracts of the selected Sudanese plants

Plants	Part used	Solvent used	% inhibition of cell proliferation			
			HCT-116	MCF-7	CCD	**HUVECs**
Indgosfera spinosa	Stem	n-Hexane	27.79	40.43	4.16	5.64
		Ethanol	14.98	42.48	5.42	6.08
		Methanol	11.16	41.86	4.93	6.79
		Water	12.95	33.90	6.18	8.11
Nicotiana gluaca	Leaves	n-Hexane	95.08	99.38	45.63	62.05
		Ethanol	49.66	81.10	38.57	36.25
		Methanol	19.83	75.58	10.15	22.64
		Water	17.20	49.81	8.09	19.79
	Stem	n-Hexane	87.07	92.01	44.35	59.46
		Ethanol	33.58	49.28	32.63	25.43
		Methanol	43.47	53.51	40.47	27.87
		Water	28.95	57.53	38.56	19.89
Tephrosia apollinea	Aerial part	n-Hexane	53.86	87.20	9.65	26.95
		Ethanol	46.91	86.94	7.88	16.28
		Methanol	47.43	87.09	10.23	18.91
		Water	16.86	25.60	4.68	5.79
Tamarix nilotica	Leaves	n-Hexane	29.33	47.25	8.77	14.58
		Ethanol	7.8	46.61	5.83	22.89
		Methanol	19.61	46.14	6.65	17.35
		Water	6.93	28.47	3.87	11.68
Combretum hartmannianum	Leaves	n-Hexane	80.03	95.76	7.75	16.62
		Ethanol	33.69	54.04	6.35	7.88
		Methanol	17.61	78.22	7.79	9.58
		Water	26.02	36.30	6.23	5.65
	Bark	n-Hexane	26.30	42.43	5.74	6.61
		Ethanol	29.11	76.79	6.48	8.36
		Methanol	13.75	73.64	9.57	11.59
		Water	28.40	49.92	10.11	9.47
Capparis decidua	Stem	n-Hexane	33.78	50.07	5.34	15.45
		Ethanol	8.5	41.99	3.69	5.77
		Methanol	14.76	39.87	5.58	19.57
		Water	21.82	54.41	6.76	4.38

Hassan *et al.*

Hassan *et al. BMC Complementary and Alternative Medicine* 2014 14:406, doi: 10.1186/1472-6882-14-406

Table 4: IC50 (µg/ml) values of the active extracts of the selected Sudanese plants

Extracts	Carcinoma cell lines		Normal cell lines	
	HCT 116	**MCF-7**	**HUVEC**	**CCD-18CO**
C. hartmannianum	27.2	8.48	116.5	466.4
N. gluaca	5.4	10.78	46.2	79.82
T. apollinea	20.2	29.36	98.4	345.6
Positive controls	5-Flourouracil	Tamoxifen	Betulinic acid	Betulinic acid
	3.9	6.67	88.4	128.9

Hassan *et al.*

Hassan *et al. BMC Complementary and Alternative Medicine* 2014 **14**:406, doi:10.1186/1472-6882-14-406

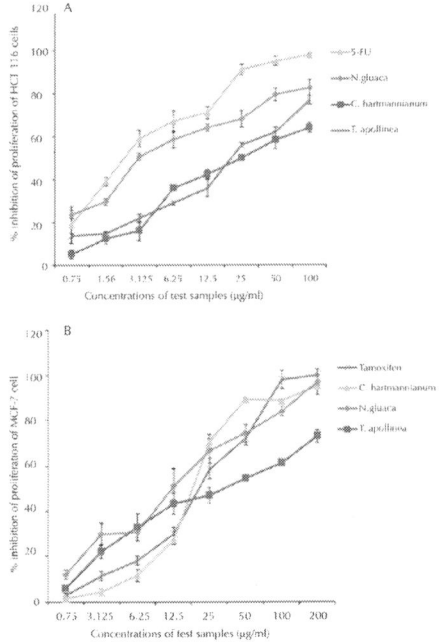

Figure 2: Dose dependent inhibitory effect of the active extracts of the selected Sudanese plants against HCT 116 (A) and MCF-7 (B) cell lines. The values are expressed as mean ± SEM (n =6).

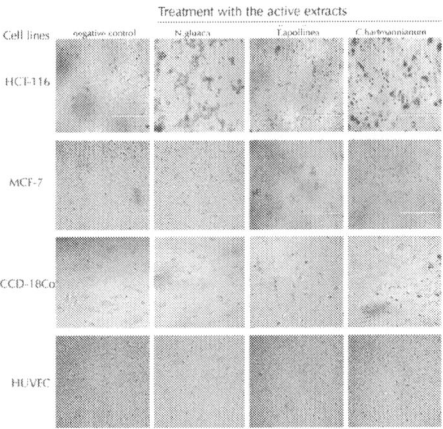

Figure 3: Photomicrographic images of the cancer (HCT 116 and MCF-7) and normal (HUVEC and CCD-18Co) cell lines taken under an inverted

phase-contrast microscope at × 200 magnification with a digital camera at 48 hours after treatment with the active extracts of the selected Sudanese plants. The controlled cells showed a fully confluent growth with the compact layer of proliferating cells. Whereas, treatment with the active extracts caused a drastic reduction in the density of cell-population however, treatment with the extracts did not produced significant cytotoxicity against the tested normal cells (HUVEC and CCD-18Co) when compared to the negative control group. The picture revealed that the cells revealed sever morphological changes in their native cellular characteristics. Treatment caused the cells to lose the psuedopodial like membrane projections. Higher magnification of the photomicrographs revealed several clear features of apoptosis, such as the membrane blebbing, nuclear condensation and apoptotic bodies in the treated cells.

Total Phenolic Contents in the Extracts

The amount of total phenolic compounds present in each extract was determined from linear regression equation of calibration curve, {y =0.0034 + 0.0144 (R^2 = 0.9976) and expressed as Gallic acid equivalent in mg/ml of extracts. Table 5 depicts the result of assessment of total phenolic contents in all the tested extracts. It is found that, hexane extracts of all the plants were deprived of total phenolic contents, either no or negligible amount of phenolic contents were detected in all hexane extracts. However, the methanol, ethanol and water extracts of the tested plants showed considerable level of phenolic contents. The results showed that, both stem bark and leaves of *C. hartmannianum* displayed high contents of total phenolics in which ethanol, methanol and water extracts showed 404.05 ± 0.06, 96.56 ± 0.05 and 523.36 ± 0.00 mg GAE/g, respectively for stem bark, whereas 169.19 ± 0.02, 392.83 ± 0.01, 268.21 ± 0.03 mg GAE/g, respectively for leaves. The results showed that methanol and water extracts of *T. nilotica* also demonstrated significant ($p < 0.05$) level of phenolic contents with 333.96 ± 0.01 and 171.66 ± 0.02 mg GAE/g, respectively. The leaves extracts of *N. glauca* showed moderate level of total phenolic contents (ethanol extract =111.62 ± 0.01 mg GAE/g; methanol extract =139.99 ± 0.01 mg GAE/g; water extract =105.66 ± 0.01 mg GAE/g).

Table 5: Correlation between antioxidant activity of different extracts of selected Sudanese plants and the total content of flavonoids and phenolics in the extracts

Plants	Part used	Solvent used	DPPH (IC50 in µg/ml)	Total flavonoids (mg/g)	Total phenolics (mg/g)
Indgosfera spinosa	Stem	n-Hexane	>1000	28.1 ± 0.00	ND
		Ethanol	192.63	40.2 ± 0.04	115.4 ± 0.01
		Methanol	140	36.6 ± 0.01	161.4 ± 0.03
		Water	>1000	13.3 ± 0.01	78.5 ± 0.001
Nicotiana gluaca	Leaves	n-Hexane	>1000	88.1 ± 0.01	ND
		Ethanol	54.76	77.1 ± 0.00	111.6 ± 0.01
		Methanol	13.437	64.5 ± 0.002	139.9 ± 0.01
		Water	30.05	57.6 ± 0.001	105.6 ± 0.01
	Stem	n-Hexane	>1000	58.6 ± 0.02	3.46 ± 0.01
		Ethanol	160.94	13.1 ± 0.01	ND
		Methanol	119.45	17.9 ± 0.01	41.02 ± 0.003
		Water	591.98	ND	12.5 ± 0.02
Tephrosia apollinea	Aerial part	n-Hexane	>1000	66.9 ± 0.01	ND
		Ethanol	120.22	78.1 ± 0.00	ND
		Methanol	48.803	10.7 ± 0.01	25.7 ± 0.00
		Water	2835.07	23.5 ± 0.02	23.5 ± 0.01
Tamarix nilotica	Leaves	n-Hexane	>1000	67.8 ± 0.00	ND
		Ethanol	880.70	13.1 ± 0.04	15.3 ± 0.04
		Methanol	12.62	24.9 ± 0.02	333.9 ± 0.01
		Water	67.53	22.6 ± 0.001	171.6 ± 0.02
Combretum hartmannianum	Leaves	n-Hexane	>1000	42. ±**0.002**	ND
		Ethanol	146.35	45.2 ± 0.001	169.2 ± 0.02
		Methanol	14.067	57.7 ± 0.001	392.8 ± 0.01
		Water	967.55	27.6 ± 0.02	268.2 ± 0.03
	Bark	n-Hexane	>1000	42.1 ± 0.01	1.9 ± 0.01
		Ethanol	9.47	28.1 ± 0.02	404.1 ± 0.06
		Methanol	28.91	24.2 ± 0.02	96.5 ± 0.05
		Water	22.47	29.7 ± 0.01	523.3 ± 0.01
Capparis decidua	Stem	n-Hexane	212.56	67.5 ± 0.02	85.2 ± 0.01
		Ethanol	900.58	8.5 ± 0.02	19.9 ± 0.01
		Methanol	76.38	4.6 ± 0.02	40.3 ± 0.2

		Water	>1000	ND	335.7 ± 0.01

ND = Not Detected. The results are presented as mean ± SEM. Each experiment was repeated three times; (n = 3).

Hassan *et al.*

Hassan *et al.* BMC *Complementary and Alternative Medicine* 2014 14:406, doi: 10.1186/1472-6882-14-406

Total Flavonoids Contents in the Extracts

The total flavonoids contents in each extract were determined from linear regression equation of calibration curve obtained from the different concentrations of quercetin [$y = 0.0032 + 0.0669$ ($R^2 = 0.986$)] and the results were expressed as mg quercetin equivalent/g of extracts (Table 5). Hexane extracts of all the tested plants and their parts showed higher content of flavonoids than compared to the other solvent extracts. However, on average all the extracts demonstrated considerably high amount of flavonoids in *N. glauca* leaves with hexane extract =88.150 ± 0.01 mg quercetin equivalent/g, ethanol extract =77.083 ± 0.00 mg quercetin equivalent/g, methanol extract =64.583 ± mg quercetin equivalent/g and water extract =57.667 ± mg quercetin equivalent/g. Extracts of aerial parts of *T. apollinea* also showed significant level of flavonoids with hexane extract =66.950 ± 0.01 mg quercetin equivalent/g, ethanol extract =78.092 ± 0.00 mg quercetin equivalent/g, methanol extract =10.708 ± 0.01 mg quercetin equivalent/g and water extract =23.575 ± 0.02 mg quercetin equivalent/g. The total flavonoid contents measured in the extracts is given in the Table 5.

DPPH Scavenging Effect of the Extracts

IC_{50} values of DPPH scavenging activity of the extracts is tabulated in Table 5. On average, the extracts prepared from methanol demonstrated the most potent antioxidant activity, whereas the extracts prepared from the solvent hexane displayed poor DPPH scavenging activity as the IC_{50} values estimated to be more than 1000 µg/ml. Among the tested plants, *C. hartmannianum* stem bark exhibited significant ($p < 0.01$)

antioxidant activity as lowest IC_{50} values were calculated for ethanol extract ($IC_{50} = 9.47$ µg/ml), methanol extract ($IC_{50} = 28.91$ µg/ml) and water extract ($IC_{50} = 22.47$ µg/ml). Similarly, *N. glauca* leaves also demonstrated significant ($p < 0.05$) antioxidant activity. The IC_{50} values for ethanol, methanol and water extracts of *N. glauca* leaves were calculated as 54.76, 13.437 and 30.05 µg/ml, respectively. The other plants displayed either moderate or insignificant DPPH scavenging activity (Table 5).

DISCUSSION

Increasing evidences in both experimental and clinical studies suggest that angiogenesis and oxidative stress play a cumulative role in the pathogenesis of malignancy. Angiogenesis, a process of formation of new blood vessel from pre-existing vessels, strongly implicates in the metastatic carcinogenesis. The role of reactive oxygen species (ROS) in tumorous angiogenesis has been extensively investigated and various connections have been established [25]. ROS can function as signaling molecules to mediate various angiogenic-related responses such as cell proliferation, differentiations and migration [26, 27].

Natural anticancer medicines discovered from various medicinal systems that have been derived from traditional knowledge and practiced in many countries. Similarly, many herbal extracts and products are traditionally being used in Sudan for the treatment of cancer [28]. However, such medicinal plants have not gained clinical importance as botanicals due to the lack of systematic and scientific evidence presented through the suitable standard experimental procedures. In the present study six indigenous anticancer medicinal plants (*I. spinosa, N. glauca, T. apollinea, T. nilotica, C. hartmannianum* and *C. decidua*) from Sudan were studied. Since the people of Sudan have long been using these plants as either food or medicine, the plants considered as an integral part of the local pharmacopoeia. In the present study, different parts of the plants were selected to prepare 32 extracts using sequential extraction method with 4 solvents of different polarity. The rationale for performing extractions from non-polar to polar solvents is to confirm and validate the bio-efficacy in the aqueous extractions performed in the traditional manner in the form of tonics and aqueous pastes. In addition, the extracts with different polarities provide an idea about the specific phytochemical groups of the active principles present in the extracts.

This study aimed to evaluate the anti-carcinogenic activities of the 6 Sudanese medicinal plants and to correlate these activities with phytochemical analysis and antioxidant capability. Two assays were used to assess the anti-carcinogenic properties of the plant extracts. One is the MTT assay which provides a simple method for determination of cell's viability via mitochondrial activity in living cells and other one is the rat aorta ring assay, which is based on the ability of the aortic wall to produce neo-vessels in bio-matrix gels after mechanical injury or angiogenic factor stimulation.

The results of rat aortic ring assay showed that, out of 6 tested plants, *T. apollinea*, *C. hartmannianum* and *T. nilotica* showed strong inhibitory effect (more than 60%). Particularly, the ethanol extract of aerial parts of *T. apollinea* demonstrated highest anti-angiogenic activity by inhibiting 100% of the aortic microvessels. Noteworthily, all the extracts of *C. hartmannianum*displayed significant anti-angiogenic activity.

Whereas, results of the cytotoxic assay showed that among all the extracts, the highly non-polar solvent extract i.e., hexane extract demonstrated higher cytotoxic activity than the other solvent extracts. Among the 6 plants, the hexane extracts of *N. glauca* leaves and stem bark exhibited most potent anti-proliferative effect on all the tested cancer cell lines. However, for the plant *T. apollinea*, both non-polar (hexane) and polar (ethanol) extracts displayed selective cytotoxicity towards hormone dependent breast cancer cell line (MCF-7). Consistently, again the extracts of *C. hartmannianum* leaves and stem bark exhibited strong inhibitory effect against the proliferation of HCT-116 and MCF-7 cells. Noteworthily, the extracts of the most effective plants (*N. glauca*, *T. apollinea* and *C. hartmannianum*) did not produce significant cytotoxic effects against the normal cell lines (HUVEC and CCD-18Co).

Altogether, the most biologically active plants which showed significant antiangiogenic as well as cytotoxic activities were *N. glauca*, *T. apollinea*, *C. hartmannianum* and *T. nilotica*. These findings were further supported by the results of DPPH scavenging activity. The results showed that, most of the extracts of *C. hartmannianum* displayed strong DPPH quenching ability with lowest IC_{50} (9 µg/ml), which was followed by *N. glauca*, *T. nilotica* and *T. apollinea* (Table 5).

The antiangiogenic and anti-proliferative effects of the plants may be due to their potential antioxidant activity, which further attributes to the collective contribution of phenolics and flavonoids present it the respective extracts. The findings of the present study revealed that these plants are enriched with phenolic and flavonoid contents than compared to the other plants which have shown less bio-efficacy. On average, the plants *N. glauca, T. apollinea, C. hartmannianum*and *T. nilotica* exhibited high antioxidant activity in DPPH free radical scavenging assay. This may support the traditional usage of these plants to improve complications such oxidative stress in cancer and other diseases. Many health beneficial effects of flavonoids are attributed to their ability to act as antioxidants. Research studies have shown that flavonoids as single electron donors can stabilize and scavenge the free radicals, which in conditions of oxidative stress may initiate angiogenesis or carcinogenesis. Similarly, phenolics have strong capability to interfere in a series of physiological events in biological systems, including those relating to oxidation processes [29].

Phenolics and flavonoids mostly found in plants are reported to have numerous biological effects including antioxidant, anti-neovascularization, antiproliferation and anticarcinogenic properties and are therefore considered for their important dietary roles as antioxidants and chemoprotective agents. Recently, intensive research has been focused on studying the naturally occurring phenolics and flavonoids that are able to decrease the generation of reactive oxygen species (ROS) in biological system. Oxidative stress contributed by ROS plays a critical role in the pathologies related with chronic disease such as cancer and excessive vascularization [30]. ROS-induced development of cancer involves malignant transformation due to DNA mutations and altered gene expression through epigenetic mechanisms which in turn leads to the uncontrolled proliferation of cancerous cells. Further, high levels of ROS are observed in various cancerous cells and a number of accumulating evidences [31-33] suggest that ROS function as key signaling molecules to stimulate various growth-related responses that eventually initiate angiogenesis and tumorigenesis [27]. Several studies demonstrated a significant role of phenolics in growth inhibition of breast, colon, prostate, ovary, and endometrium and lung cancer cells [34, 35].

The present study confirmed that the extracts of *N. glauca, T. apollinea* and *C. hartmannianum*demonstrated selective cytotoxicity towards human breast and colon cancer cell lines while being less cytotoxic against the normal cells. Such selective cytotoxic activity suggested that the active substances interact with special cancer-associated receptors or cancer cell special molecule, thus triggering some mechanisms that cause cancer cell death [36]. In observation under EVOS f1 digital microscope, typical apoptotic characteristics were observed, including cell membrane blebbing, loss of pseudopodia-like cellular projections, nuclear condensation, and separated apoptotic bodies (Figure 3). In addition, the treated MCF-7 cells displayed typical signs of apoptosis such as shrinkage of cells, chromatin condensation and crescent shaped nuclei (Figure 3, MCF-7 row). Several reports have revealed that flavonoids are able to inhibit the growth of cancer cells *in vitro* [37-39]. These findings are confirmed by several *in vivo* studies [40]. Flavonoids exert their anticancer action through affecting key mechanisms involved in cancer pathogenesis. Flavonoids are effective antioxidant and antiangiogenic agents. In initial stages, they inhibit metabolic activation of carcinogens. In progression phases they induce apoptosis, inhibit angiogenesis, cancer cell proliferation and tumor metastasis [41].

The magnitude of phenolic compounds in redox system acts either as reducing agents, hydrogen donators, and singlet oxygen quenchers or in some cases as a metal chelating agents [42] thereby, the free radicals which generated in the metabolic pathways could be neutralized by phenolic compounds [43]. From the results obtained in this study, there is an obvious correlation between antiangiogenic and antioxidant activity, since the polyphenols inhibit the initiation and progression of angiogenesis [44-46]. Therefore, plant polyphenols may play an important role in halting angiogenesis, as well as have ability to prevent oxidant potentials of free radicals as natural source of antioxidants.

Results of the present study were completely in agreement with the study reported by Mariod et al., [47] on the various extracts of *C. hartmannianum* for its antioxidant property and total phenolic content. In the present study, *C. hartmannianum* demonstrated strong antiproliferative and antiangiogenic activities. It is reported that *C. hartmannianum* has strong capability to inhibit tyrosine kinase [48]. Tyrosine kinase is an important cellular signaling protein which has essential and critical role in several biological activities including cell proliferation and angiogenesis [49]. Several tyrosine kinase inhibitors such as, bevacizumab, sunitinib, sorafenib and pazopanib were recently approved for treatment of patients with malignant carcinoma [50, 51]. In addition, many other anti-angiogenic tyrosine kinase inhibitors are being studied in phase I-III clinical trials in order to validate their beneficial anti-tumor, anti-metastatic and anti-angiogenic activities in human beings [52].

In this study, another herb that has shown a promising cytotoxicity on the tested cancer cell lines was *N. glauca*. Ibrahim and El-Sharkawy [53] reported the antioxidant activity and presence of phenolics and flavonoids in the leaves of *N. glauca* however, the present study reported that even stem bark extracts of the plant has remarkable antioxidant activity which could be attributed to the presence of high levels of phenolics and flavonoids in it.

The present study showed that both non-polar and polar extracts of aerial part of *T. apollinea* (Figure 3) exhibited selective antiproliferative effect against breast cancer cell line MCF-7. However, in our previous study [54] we reported the isolation of (-)-pseudosemiglabrin from aerial parts of *T. apollinea* and its antiproliferative activity on prostate, leukemia, breast and colon cancer cells. In the present study, the anticancer and antiangiogenic activities of nonpolar (hexane) extracts could be partly attributed to the high levels of flavonoid. It is well known that the plant-flavonoids demonstrated promising anticancer and antiangiogenic activities [55]. Several studies [56] reported the presence of complex prenylated flavones and phenolic compounds in *T. apollinea* and thus the antiangiogenic and antiproliferative properties of the herb recorded in the present study can be attributed to the cumulative effects of such bioactive constituents.

For the first time the present study reported the cytotoxic potentials of leaves of *T. nilotica* against human colon (HCT-116) and breast (MCF-7) cancer cells. However, the earlier reports mentioned that *T. nilotica* has selective cytotoxic potential against liver cell carcinoma (Huh-7), while being non-toxic to other cancer cells [57]. Nevertheless, the results of the present study agree in part with the previous findings [57] in the sense that *T. nilotica* has noticeable DPPH quenching capability. These findings are in agreement with some earlier reports [58,59] wherein, various types of phenolic constituents such as phenolic glyceride, phenolic lactone, phenolic aldehydes and dimeric phenols were isolated from the of *T. nilotica*.

Anti-angiogenic agent could target the cancer or endothelial cells at any of the steps necessary for carcinogenesis or neovascularization, such as proliferation, differentiation, migration, or tube formation [60]. Angiogenesis inhibitors act by actively promoting apoptosis in cells. *In vitro* and *in vivo* investigations have proved that many endogenous antiangiogenic compounds induce cytotoxicity via apoptotic cellular death [61].

The results of the present study could be very helpful as preliminary data in the search for new antitumor compounds from the tested Sudanese traditional medicinal plants. These plants have the potential to be chemically standardized and used as herbal medicines or developed into pharmaceutical drugs for the treatment of angiogenesis-dependent human ailments such as cancer and other hyperproliferative disorders. Bioassay-guided phytochemical and pharmacological studies are under investigation in an attempt to isolate and characterize the active constituents from the extracts of the plants which have shown promising antiangiogenic and anticancer properties.

CONCLUSIONS

In conclusion the present study revealed that among the six Sudanese medicinal plants tested, *N. glauca*, *T. apollinea*, *C. hartmannianum* and *T. nilotica* found to be most biologically effective herbs with significant antiangiogenic and antineoplastic effects. The biological activities observed in the study could be attributed to the antioxidant and higher phenolic and flavonoid contents of the plants. Furthermore, the extracts of these plants displayed either negligible or insignificant cytotoxicity against the tested human normal cells. Thus, these herbs could be considered as promising candidates for the development of novel chemopreventive or chemotherapeutic formulations with reduced side effects. The results obtained in the present study justify the traditional use of these medicinal plants to cure neoplasia. However, further studies to isolate the active compounds and to investigate the mode of action using *in vivo* xenograft experimental models are warranted against pathological neovascularization and tumor malignancy.

AUTHORS' CONTRIBUTIONS

AMSAM, MBKA, and LEAH designed the experiments. LEAH, HB and NSM carried out the phytochemical analysis. LEAH, NSM and ZDN performed the *ex vivo* antiangiogenic activity. LEAH, ZGN and HB participated in cytotoxic study on the plants. LEAH, MBKA, ASAM and AMSAM analyzed the data and interpreted the results. LEAH, MBKA and ASAM drafted the manuscript. All the authors read and approved the final manuscript.

ACKNOWLEDGMENTS

We wish to acknowledge TWAS-USM (FR number: 3240240313) for the fellowship to LEAH and financial support for this research. In addition, the authors wish to acknowledge the support from Universiti Sains Malaysia for the research study through Research University Team (RUT) Grant No.: 1001/PFARMASI/851001.

REFERENCES

1. Bray F, Ren JS, Masuyer E, Ferlay J: Global estimates of cancer prevalence for 27 sites in the adult population in 2008. *Int J Cancer* 2013, 132(5):1133-1145.

2. Chorawala MR, Oza PM, Shah GB: Mechanisms of anticancer drugs resistance: an overview. *Int J Pharma Sci Drug Res* 2012, 1:01-09.

3. Cragg GM, Newman DJ: Plants as a source of anti-cancer agents. *J Ethnopharmacol* 2005, 100(1–2):72-79.

4. Jung Park E, Pezzuto J: Botanicals in cancer chemoprevention. *Cancer Metastasis Rev* 2002, 21(3–4):231-255.

5. Shoeb M: *Anticancer Agents from Medicinal Plants*. Pharmacology: Bangladesh Journal of; 2006:1(2006).

6. Newman DJ, Cragg GM: Natural products as sources of new drugs over the 30 years from 1981 to 2010. *J Nat Prod* 2012, 75(3):311 335.

7. Jaspars M, Lawton LA: Cyanobacteria - a novel source of pharmaceuticals. *Curr Opin Drug Discov Devel* 1998, 1(1):77-84.

8. Folkman J: What is the evidence that tumors are angiogenesis dependent? *J Natl Cancer Inst* 1990, 82(1):4-6.

9. Folkman J: Role of angiogenesis in tumor growth and metastasis. *Semin Oncol* 2002, 29(6, Supplement 16):15-18.

10. Scappaticci FA: The therapeutic potential of novel antiangiogenic therapies. *Expert Opin Investig Drugs* 2003, 12(6):923-932.

11. Dell'Eva R, Pfeffer U, Vené R, Anfosso L, Forlani A, Albini A, Efferth T: Inhibition of angiogenesis in vivo and growth of Kaposi's sarcoma xenograft tumors by the anti-malarial artesunate. *Biochem Pharmacol* 2004, 68(12):2359-2366.

12. Hurwitz H, Fehrenbacher L, Novotny W, Cartwright T, Hainsworth J, Heim W, Berlin J, Baron A, Griffing S, Holmgren E, Ferrara N, Fyfe G, Rogers B, Ross R, Kabbinavar F:Bevacizumab plus Irinotecan, Fluorouracil, and Leucovorin for Metastatic Colorectal Cancer. *N Engl J Med* 2004, 350(23):2335-2342.

13. Al-Suede FSR, Farsi E, Ahamed MKB, Ismail Z, Abdul Majid AS, Abdul Majid AMS: Marked antitumor activity of cat's whiskers tea (Orthosiphon stamineus) extract in orthotopic model of human colon tumor in nude mice. *J Biochem Tech* 2012, 5(28 December 2012):S170-S176.

14. Dawidowicz AL, Wianowska D, Baraniak B: The antioxidant properties of alcoholic extracts from Sambucus nigra L. (antioxidant properties of extracts). *LWT Food Sci Technol* 2006, 39(3):308-315.

15. Yoysungnoen P, Wirachwong P, Changtam C, Suksamrarn A, Patumraj S: Anti-cancer and anti-angiogenic effects of curcumin and tetrahydrocurcumin on implanted hepatocellular carcinoma in nude mice. *World J Gastroenterol* 2008, 14(13):2003 2009.

16. Munzel T, Gori T, Bruno RM, Taddei S: Is oxidative stress a therapeutic target in cardiovascular disease? *Eur Heart J* 2010, 31(22):2741-2748.

17. Dell'Agli M, Buscialà A, Bosisio E: Vascular effects of wine polyphenols. *Cardiovasc Res* 2004, 63(4):593-602.

18. Galal M, Bashir AK, Salih AM, Adam SL: Activity of water extracts of Albizia Anthelmintica and A. lebek barks against experimenta Hymenolepis diminuta infection in rats. *J Ethnopharmacol* 1991, 33(1991):333-337.

19. Eatock MM, Schatzlein A, Kaye SB: Tumour vasculature as a target for anticancer therapy. *Cancer Treat Rev* 2000, 26(3):191-204.

20. Al-Salahi OSA, Kit-Lam C, Majid AM, Al-Suede FS, Mohammed Saghir SA, Abdullah WZ, Ahamed MB, Yusoff NM: Anti-angiogenic quassinoid-rich fraction from Eurycoma longifolia modulates endothelial cell function. *Microvasc Res* 2013, 90:30-39.

21. Ahamed MB, Aisha AF, Nassar ZD, Siddiqui JM, Ismail Z, Omari SM, Parish CR, Majid AM:Cat's whiskers tea (Orthosiphon stamineus) extract inhibits growth of colon tumor in nude mice and angiogenesis in endothelial cells via suppressing VEGFR phosphorylation. *Nutr Cancer* 2012, 64(1):89-99.

22. Al-Suede FSR, Khadeer Ahamed MB, Abdul Majid AS, Baharetha HM, Hassan LE, Kadir MOA, Nassar ZD, Abdul Majid AMS: Optimization of cat's whiskers tea (orthosiphon stamineus) using supercritical carbon dioxide and selective chemotherapeutic potential against prostate cancer cells. *Evid Based Complement Alternat Med* 2014, 2014:15.

23. Baharetha HM, Nassar ZD, Aisha AF, Khadeer AMB, Al-Suede FSR, Abd KMO, Zhari I, Abdul MAMS: Proapoptotic and antimetastatic properties of supercritical CO2 extract of Nigella sativa Linn. against breast cancer cells. *J Med Food* 2013, 16(12):1121-1130.

24. Khadeer Ahamed MB, Krishna V, Dandin CJ: In vitro antioxidant and in vivo prophylactic effects of two gamma-lactones isolated from Grewia tiliaefolia against hepatotoxicity in carbon tetrachloride intoxicated rats. *Eur J Pharmacol* 2010, 631(1–3):42-52.

25. Gibellini L, Pinti M, Nasi M, De Biasi S, Roat E, Bertoncelli L, Cossarizza A: Interfering with ROS metabolism in cancer cells: the potential role of quercetin. *Cancer* 2010, 2(14 June 2010):1288-1311.

26. Coso S, Harrison I, Harrison CB, Vinh A, Sobey CG, Drummond GR, Williams ED, Selemidis S: NADPH oxidases as regulators

of tumor angiogenesis: current and emerging concepts. *Antioxid Redox Signal* 2012, 16(11):1229-1247.

27. Ushio-Fukai M, Nakamura Y: Reactive oxygen species and angiogenesis: NADPH oxidase as target for cancer therapy. *Cancer Lett* 2008, 266(1):37-52.

28. Hilmi Y, Abushama MF, Abdalgadir H, Khalid A, Khalid H: A study of antioxidant activity, enzymatic inhibition and in vitro toxicity of selected traditional sudanese plants with anti-diabetic potential. *BMC Complement Altern Med* 2014, 14(1):1472-6882.

29. Balasundram N, Sundram K, Samman S: Phenolic compounds in plants and agri-industrial by-products: antioxidant activity, occurrence, and potential uses. *Food Chem* 2006, 99(1):191-203.

30. Kampa M, Nifli AP, Notas G, Castanas E: Polyphenols and cancer cell growth. *Rev Physiol Biochem Pharmacol* 2007, 159:79 113.

31. Yasuda M, Ohzeki Y, Shimizu S, Naito S, Ohtsuru A, Yamamoto T, Kuroiwa Y: Stimulation of in vitro angiogenesis by hydrogen peroxide and the relation with ETS-1 in endothelial cells. *Life Sci* 1999, 64(4):249-258.

32. Yeldandi AV, Rao MS, Reddy JK: Hydrogen peroxide generation in peroxisome proliferator-induced oncogenesis. *Mutat Res* 2000, 448(2):159-177.

33. Irani K, Xia Y, Zweier JL, Sollott SJ, Der CJ, Fearon ER, Sundaresan M, Finkel T, Goldschmidt-Clermont PJ: Mitogenic signaling mediated by oxidants in Ras-transformed fibroblasts. *Science* 1997, 275(5306):1649-1652.

34. Galati G, O'Brien PJ: Potential toxicity of flavonoids and other dietary phenolics: significance for their chemopreventive and anticancer properties. *Free Radic Biol Med* 2004, 37(3):287-303.

35. Baghel SS, Shrivastava N, RS Baghel PA, Rajput S: A review of quercetin: antioxidant and anticancer properties. *World J Pharm Pharmaceutical Sci* 2012, 1(1):146-160.

36. Harada H, Noro T, Kamei Y: Selective antitumor activity in vitro from marine algae from Japan coasts. *Biol Pharm Bull* 1997, 20(5):541-546.

37. Moghaddam G, Ebrahimi SA, Rahbar-Roshandel N, Foroumadi A: Antiproliferative activity of flavonoids: influence of the sequential

methoxylation state of the flavonoid structure. *Phytother Res* 2012, 26(7):1023-1028.

38. Delmulle L, Bellahcène A, Dhooge W, Comhaire F, Roelens F, Huvaere K, Heyerick A, Castronovo V, De Keukeleire D: Anti-proliferative properties of prenylated flavonoids from hops (Humulus lupulus L.) in human prostate cancer cell lines. *Phytomedicine* 2006, 13(9–10):732-734.

39. Chidambara Murthy KN, Kim J, Vikram A, Patil BS: Differential inhibition of human colon cancer cells by structurally similar flavonoids of citrus. *Food Chem* 2012, 132(1):27-34.

40. Batra P, Sharma A: Anti-cancer potential of flavonoids: recent trends and future perspectives. *3 Biotech* 2013, 3(6):439-459.

41. Clere N, Faure S, Martinez MC, Andriantsitohaina R: Anticancer properties of flavonoids: roles in various stages of carcinogenesis. *Cardiovasc Hematol Agents Med Chem* 2011, 9(2):62-77.

42. Gordon MH: The Mechanism of Antioxidant Action in Vitro. In *Food Antioxidants*. Edited by Hudson BJF. Netherlands: Springer; 1990:1-18.

43. Miliauskas G, Venskutonis PR, van Beek TA: Screening of radical scavenging activity of some medicinal and aromatic plant extracts. *Food Chem* 2004, 85(2):231-237.

44. Stoclet J-C, Chataigneau T, Ndiaye M, Oak MH, El Bedoui J, Chataigneau M, Schini-Kerth VB: Vascular protection by dietary polyphenols. *Eur J Pharmacol* 2004, 500(1–3):299-313.

45. Oak MH, El Bedoui J, Schini-Kerth VB: Antiangiogenic properties of natural polyphenols from red wine and green tea. *J Nutr Biochem* 2005, 16(1):1-8.

46. Walter A, Etienne-Selloum N, Brasse D, Schleiffer R, Bekaert V, Vanhoutte PM, Beretz A, Schini-Kerth VB: Red wine polyphenols prevent acceleration of neovascularization by angiotensin II in the ischemic rat hindlimb. *J Pharmacol Exp Ther* 2009, 329(no. 2):329-699.

47. Mariod A, Matthäus B, Hussein IH: Antioxidant activities of extracts from Combretum hartmannianum and Guiera senegalensis on the oxidative stability of sunflower oil. *Emir J Food Agric* 2006, 18(2006):20-28.

48. Ali H, König GM, Khalid SA, Wright AD, Kaminsky R: Evaluation of selected Sudanese medicinal plants for their in vitro activity against hemoflagellates, selected bacteria, HIV-1-RT and tyrosine kinase inhibitory, and for cytotoxicity. *J Ethnopharmacol* 2002, 83(3):219-228.

49. Gotink KJ, Verheul HM: Anti-angiogenic tyrosine kinase inhibitors: what is their mechanism of action? *Angiogenesis* 2010, 13(1):1-14.

50. Faivre S, Demetri G, Sargent W, Raymond E: Molecular basis for sunitinib efficacy and future clinical development. *Nat Rev Drug Discov* 2007, 6(9):734-745.

51. Wilhelm S, Carter C, Lynch M, Lowinger T, Dumas J, Smith RA, Schwartz B, Simantov R, Kelley S: Discovery and development of sorafenib: a multikinase inhibitor for treating cancer. *Nat Rev Drug Discov* 2006, 5(10):835-844.

52. Patel PH, Chaganti RS, Motzer RJ: Targeted therapy for metastatic renal cell carcinoma. *Br J Cancer* 2006, 94(5):614 619.

53. Ibrahim B, Saleh H, A.E.-S: *Phytochemical Investigation of Nicotiana glauca and Microbial Degradation of Nicotine as a Water Pollutant of Tobacco*. Fairford, GLO, United Kingdom: LAP LAMBERT Academic Publishing; 2012.

54. Hassan LEA, Majid ASA, Iqbal MA, Al Suede FSR, Haque RA, Ismail Z, Ein OC, Majid AMSA, M.B.K.A: Crystal structure elucidation and anticancer studies of (-)-pseudosemiglabrin: a flavanone isolated from the aerial parts of tephrosia apollinea. *PLoS ONE* 2014, 9(3):90806.

55. Dai Z-J, Lu WF, Gao J, Kang HF, Ma YG, Zhang SQ, Diao Y, Lin S, Xi-Jing W, Wu WY: Anti-angiogenic effect of the total flavonoids in Scutellaria barbata D. Don. *BMC Complement Altern Med* 2013, 13(1):1-10.

56. Abou-Douh AM, Toscano RA, Nariman N, El-Khrisy E, I.C: Prenylated flavonoids from the root of Egyptian Tephrosia apollinea–crystal structure analysis. *Z Naturforsch* 2005, B 60:458-470.

57. Abouzid S, Sleem A: Hepatoprotective and antioxidant activities of Tamarix nilotica flowers. *Pharm Biol* 2011, 49(4):392-395.

58. Nawwar MAM, Buddrus J, Bauer H: Dimeric phenolic constituents from the roots of Tamarix nilotica. *Phytochemistry* 1982, 21(7):1755-1758.

59. Barakat HH, Nawwar MA, Buddrus J, Linscheid M: Niloticol, a phenolic glyceride and two phenolic aldehydes from the roots of Tamarix nilotica. *Phytochemistry* 1987, 26(6):1837-1838.

60. Folkman J: Angiogenesis and apoptosis. *Semin Cancer Biol* 2003, 13(2):159-167.

61. Tiwari M: Apoptosis, angiogenesis and cancer therapies. *J Cancer Ther Res* 2012, 1(1):3.

Evaluation of Antioxidant Efficacy of Natural Plant Extracts against Synthetic Antioxidants in Sunflower Oil

Shweta U. Tavasalkar[1], H. N. Mishra[2], and Sanjith Madhavan[3]

[1]Food Engg. Deapartment, IIT Kharagpur, Maharashtra, India

[2]Professor, Agriculture and Food Engg. Deapartment, IIT Kharagpur, Kharagpur, West Bengal, India

[3]Synthite Industries Ltd, Kolenchery, Cochin, Kerala, India

ABSTRACT

With the improvement in living conditions, consumers increasingly reject food prepared with additives of chemical origin due to health issues. Four natural antioxidant blends prepared with hexane and acetone extract of rosemary, green tea and other additives having synergistic effect were chosen viz. OF001, OF002, OF003 and OF004 for the present study. HPLC analysis of these extracts showed that OF001 contained 4.789% carnosic acid. Similarly, OF002, OF003 and OF004 contained 3.477%, 4.676% and 7.074% carnosic acid respectively. Total phenolic content of OF001, OF002, OF003 and OF004 were found as 89.64, 89.64, 81.36, 28.29 and 77.68 mg GAE/g, respectively. Control sunflower oil found more susceptible to oxidative deterioration as peroxide value of control sunflower found to increase rapidly after 15 days of storage at ambient temperature. For sunflower oil, OF002@ 0.1% was found effective to restrict the peroxide value below threshold level up to 120 days of storage.

INTRODUCTION

The oils and fats of commerce are mixtures of lipids. They are mainly triacylglycerols (generally >95%) accompanied by diacylglycerols, mono-acylglycerols and free fatty acids, but they may also contain phospholipids, free sterols and sterol esters, tocols (tocopherols and tocotrienols), triterpene alcohols, hydrocarbons and fat-soluble vitamins. During refining, some of the minor components are removed, wholly or in part, and useful materials may be recovered [1]. Sunflower oil is one of the most popular vegetable oils and in some countries it is preferred to soybean, cottonseed and palm oils. The sunflower is the fourth largest oil source in the world, after soybean, palm and canola oil. Demand for sunflower oil increased sharply in the mid-eighties when high polyunsaturated fatty acid (PUFA) margarine became the desired table margarine for health reasons. Traditional sunflower oil is excellent for cooking, making salad dressing, margarine, and so on, but it cannot be used for manufacturing shelf-stable fried foods because of its poor oxidative stability [1]. Fats, oils and lipid-based foods deteriorate through several degradation reactions both on heating and on long term storage. The main deterioration processes are oxidation reactions and the decomposition of oxidation products which result in decreased nutritional value and sensory quality. The retardation of these oxidation processes is important for the food producer and, indeed, for all persons involved in the entire food chain from the factory to the consumer. Oxidation may be inhibited by various methods including prevention of oxygen access, use of lower temperature, inactivation of enzymes catalyzing oxidation, reduction of oxygen pressure, addition of chelating agent and the use of suitable packaging [2]. Another method of protection against oxidation is to use specific additives which inhibit oxidation. These are correctly called oxidation inhibitors, but nowadays are mostly called antioxidants. According to Halliwell and Gutteridge, the term antioxidant refers to "a substance significantly delays or prevents oxidation of that substrate". Antioxidants delay the development of off-flavors by extending the induction period. Addition of antioxidants after the end of this period tends to be ineffective in retarding rancidity development [2]. There has been an increasing interest in the use of natural antioxidants, such as tocopherols, flavonoids and rosemary (Rosmarinus officinalis L.) extracts for the preservation of food materials in recent years [3-6], because these natural antioxidants avoid the toxicity problems which may arise from the use of synthetic antioxidants, such as butylated hydroxy anisole (BHA), butylated hydroxy toluene (BHT) and propyl gallate (PG) [7,8]. Reports revealing that BHA and BHT could be toxic, and the higher manufacturing costs and lower efficiency of natural antioxidants such as tocopherols, together with the increasing consciousness of consumers with regard to food additive safety, has created a need for identifying alternative natural and probably safer sources of food antioxidants [9,10]. Natural antioxidants are more ideal as food additives, not only for their free radical scavenging properties, but also on the belief that natural products are healthier and safer than synthetic ones; thus, they are more readily acceptable to the modern consumers [11]. There is at present increasing interest both in the industry and in scientific community for spices and aromatic herbs because of their strong antioxidant and antimicrobial properties, which exceed many currently used natural and synthetic antioxidants. These properties are due to many substances, including some vitamins, flavonoids, terpenoids, carotenoids, phytoestrogens, minerals, etc. which render spices and some herbs or their antioxidant components for preservative properties in food. There has been increasing research in the role of herbs and spices as natural preservatives. Herbs and spices are not just valuable in adding flavor to foods, but their antioxidant activity also helps to preserve foods from oxidative deterioration thereby increasing their shelf life. As an example, ground black pepper has been found to reduce the lipid oxidation of cooked pork [12]. The present was therefore, undertaken in line of the above points, to study the effect of natural antioxidant blends on the oxidative stability of sunflower oil.

MATERIALS AND METHODS

Materials

Sunflower oil treated with natural antioxidant blends viz. OF001, OF002, OF003 and OF004, two synthetic antioxidants viz. tertiary butyl hydroquinone (TBHQ) and butylated hydroxyl anisole (BHA) and control sample i.e., without any antioxidant were stored at ambient temperature for 180 days and at 50°C for 90 days. Tests at 50°C were carried out in standard laboratory oven. Samples were withdrawn at 15 days interval and tested for quality parameters viz. peroxide value, free fatty acids, p-anisidine value and polar compounds. All experiments were carried out in triplicates.

Preparation and Analysis of Natural Antioxidant Blends

The natural antioxidant blends were prepared by using rosemary, green tea and other additives with different concentrations. Rosemary Ooty variety (aqueous acetone extract) and rosemary Moroccan variety (pure hexane extract) and spray dried green tea were used for preparing final testing samples viz. OF001, OF002, OF003 and OF004. Lecithin and citric acid are used for synergistic effect. Carnosic acid and carnosol content determination was carried out by using method followed by [13]. In this method, HPLC column, Prodigy ODS-2 150 x 4.6 mm, 5 micron was used as a stationary phase and mixture of acetonitrile (0.1%) and phosphoric acid (60:40) was used as a mobile phase at the rate of 1 ml/min. The detection was carried out at 230 nm. The standards of known concentration of carnosic acid (49.69%) and carnosol (2.42%) were used for preparing standard curve.

Total Phenolic Content

The total phenol content of plant extracts was determined using Folin–Ciocalteu reagent (FCR) method according to the procedure reported by [14] with some modifications. For measuring total phenolic content, gallic acid standard solution in different concentrations (5-80 mg / l) were prepared and standard curve of total phenolic content in terms of gallic acid equivalents (GAE) was prepared. 0.05g - 0.1g of sample was weighed and dissolved in alcohol and volume was made up to 25 ml. From this sample solution, 1.8 ml of solution was transferred in another volumetric flask to which 3.6 ml of 10% aqueous FCR and 14.6 ml of 100 mM sodium carbonate solution was added. The samples were kept for 2 hrs at room temperature for incubation. After 2 hrs, absorbance was measured against reagent blank (alcohol) at 765 nm using 2010 UV-IS Spectrophotometer.

Oxidative Stability of Sunflower Oil

The American Oil Chemists Society standard methods (Official Methods and Recommended Practices, 1994) were used for free fatty acids (Ca 5a-40), peroxide value (Cd 8- 53), and p-anisidine value (Cd 18-90). Polar compounds present in oil and fats were measured by IUPAC standard method 2.507. The oil stability index was determined using the Metrohm rancimat (Metrohm 743).

Statistical Analysis

All experiments were conducted in triplicates and results were represented as mean ± standard deviation. Curve fitting was done and regression coefficient of the equation for best fitted curve was calculated by using Microsoft Excel 10.

RESULT AND DISCUSSION

Analysis of Natural Antioxidant Blends

Hexane and aqueous acetone extract of rosemary and other additives having synergistic effect were used to make blends of natural antioxidant. Four natural antioxidant blends were chosen viz. OF001, OF002, OF003 and OF004 for the present study based on induction time given by accelerated study using rancimat. Carnosic acid (%), carnosol (%) and total phenolic content (mg GAE/g) of OF001, OF002, OF003 and OF004 has been listed in table 1.

Table 1: Analysis of natural plant extract

Treatment	Carnosic acid (%)	Carnosol (%)	Total phenolic content (mg GAE/g sample)
OF001	4.789	0.3094	89.64270108
OF002	3.477	0.270	81.3586176
OF003	4.676	0.689	28.2863447
OF004	7.074	0.578	77.68417203

Oil Stability Index

The oil stability index (OSI) was determined using the Metrohm Rancimat (Metrohm 743). Figure 1 set out the accelerated study of sunflower oil, which is displayed in terms of induction period i.e., the time, elapsed between starting of the oxidation process and occurrence of the secondary reaction products. Table 2 shows the induction time (h) of sunflower oil at 100°C. From table 2, it can be seen that OF001 and OF002 showed oil stability index greater than TBHQ for sunflower oil at 0.1% dosage. Also OF003 and OF004 treatments gave better OSI at 0.1% dosage than BHA.

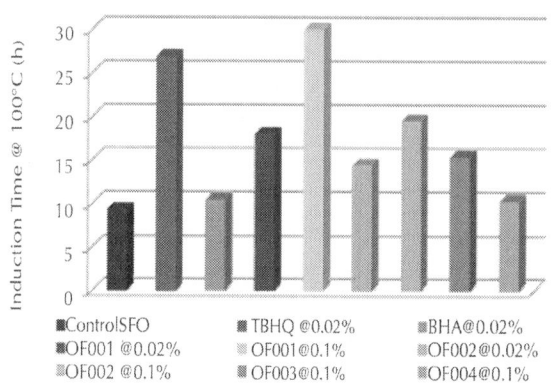

Figure 1: Effect of antioxidants on induction time of sunflower oil at 100°C.

Table 2: Accelerated study using Rancimat

Treatment	Induction Time (h)	Oil Stability Index
Control	9.32	1
TBHQ @ 0.02%	26.82	2.877
BHA @ 0.02%	10.45	1.121
OF001 @ 0.02%	18.03	1.935
OF001 @ 0.1%	29.92	3.210
OF002 @ 0.02%	14.38	1.543
OF002 @ 0.1%	19.56	2.099
OF003@ 0.1%	15.38	1.650
OF004@ 0.1%	10.32	1.107

Effect on Peroxide Value

Peroxide value (PV) is a measure of the concentration of peroxides and hydroperoxides formed during the initial stages of lipid oxidation. Peroxide value is one of the most widely used tests for oxidative rancidity in oils and fats. For this, oxidation degree on sunflower oil samples was determined by measuring PV in the absence and presence on antioxidants at ambient temperature for 180 days and at 50°C for 90 days. The influence of different antioxidants on PV of sunflower oil samples is shown in Figure 2. From figure 2, it is clear that control sunflower oil has crossed the threshold limit of PV according to CODEX Standards for edible fats and oils (10 meq. of peroxide/kg oil) after 15 days of storage. Also, control sunflower oil has reached the maximum PV (92.9092 ± 0.02464 meq. of peroxide/kg oil) after 180 days of storage at ambient temperature. From figure 3, it can be seen that there is linear increase in PV of control sunflower oil after 15 days. Also, incorporation of BHA and OF004 were not found to be effective in retarding the oxidation of sunflower oil. On the other hand, sunflower oil treated with OF001 and OF002 has shown better results than control sunflower oil even after 120 days of storage in inhibiting the oxidation of sunflower oil at ambient temperature and TBHQ was found to reach threshold limit on day 180. At 50°C, control sample attained its peak value (50.8615 ± 0.104838) on 60th day of storage as shown figure 3. The figure 3 set out the effect of antioxidants on peroxide value of sunflower oil stored at 50°C temperature. Figure 3 has shown that, after day 60, there was decrease in PV. This is due to the commencement of secondary oxidation products which were measured by p-anisidine value. The sinusoidal nature of PV curve was observed for all the treatments for sunflower oil at 50°C. This may occur at elevated temperature as peroxides are tend to form and degrade at higher temperature.

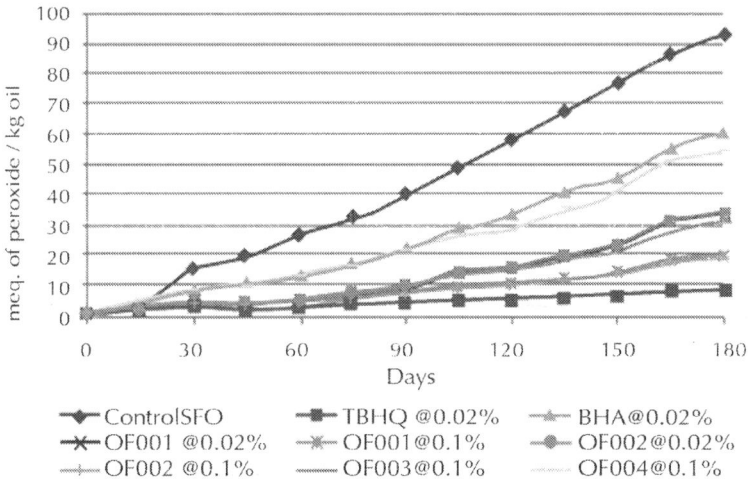

Figure 2: Effect of antioxidant type and concentration on peroxide value of sunflower oil at ambient temperature.

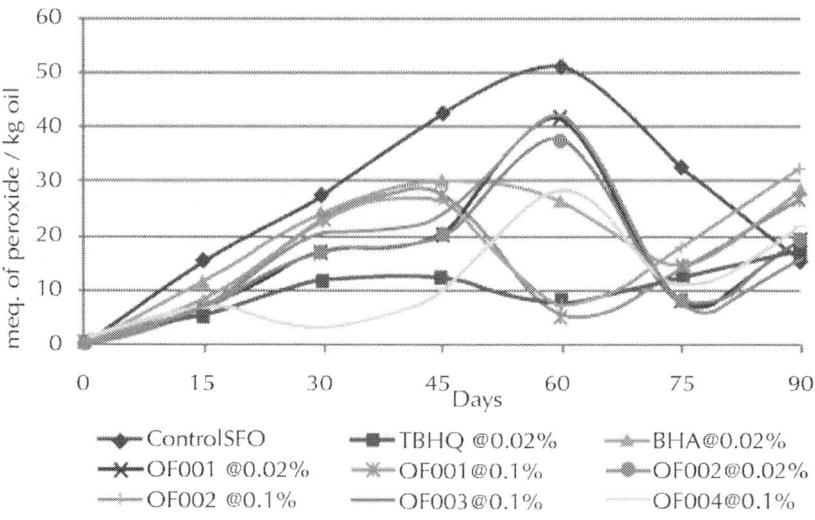

Figure 3: Effect of antioxidant type and concentration on peroxide value of sunflower oil at 50°C.

Effect on P-Anisidine Value

P-Anisidine value (AV) plays an important role in the oxidation process of edible oil and edible fats. It measures the secondary oxidation products produced during the oxidative degradation of oil, was determined by reacting p-anisidine with the oil in iso-octane and measuring the resultant color at 350 nm. This test has an enhanced sensitivity for unsaturated aldehydes, especially 2,4-dienals, but does not measure the ketonic secondary products of oxidation. The influence of different natural antioxidants treatments on p-anisidine value of sunflower oil during storage at ambient and 50°C temperature are given in figure 4 and figure 5 respectively. Control sunflower oil reached the maximum value (12.072 ± 0.0044) at ambient temperature on day 180. Sunflower oil treated with OF001 @ 0.1% showed minimum increase in AV which reached upto 10.091 ± 0.0042 from initial AV of 8.851 ± 0.0059. Also sunflower oil treated with OF002 @ 0.1% showed lesser increase in AV (10.767 ± 0.0023) as compared to the sunflower oil treated with TBHQ (10.285 ± 0.0025) and BHA (10.93 ± 0.0034).

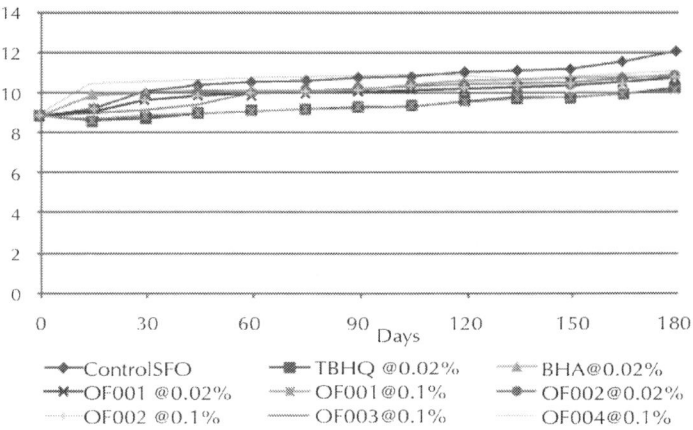

Figure 4: Effect of antioxidant type and concentration on p-anisidine value of sunflower oil at ambient temperature.

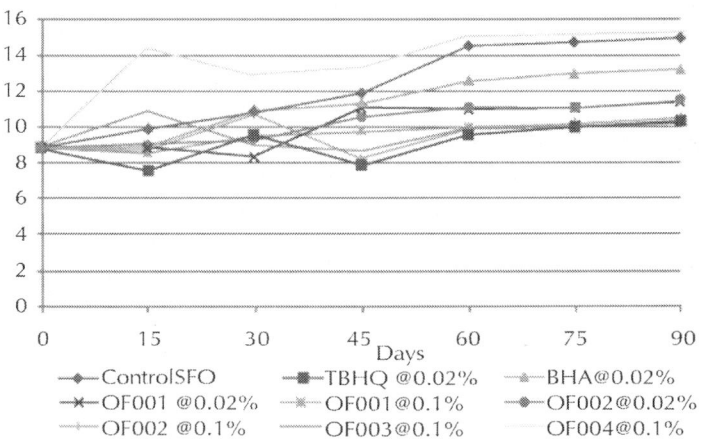

Figure 5: Effect of antioxidant type and concentration on p-anisidine value of sunflower oil at 50°C.

As shown in figure 5, it was seen that at 50°C, control sunflower reached the maximum AV up to 14.897 ± 0.039668 from initial value of 8.851 ± 0.005926. Here also at 50°C, sunflower oil treated with OF001 @ 0.1% showed minimum increase in AV up to 10.312 ± 0.012577. Sunflower oil treated with OF002 @ 0.1% showed lesser increase in AV (10.424 ± 0.010829) as compared to sunflower oil treated with TBHQ (10.305 ± 0.143942) and BHA (13.19 ± 0.011942).

Effect on Free Fatty Acid

Free fatty acid (FFA) is an important fat quality indicator during each stage of fats and oils processing. It is a measure of deodorizer efficiency and a process control tool for other processes. In crude fat, FFA or acid value estimates the amount of oil that will be lost during refining steps designed to remove fatty acids. In refined fats, a high acidity level means a poorly refined fat or fat breakdown after storage or use. As shown in figure 6, control sunflower showed the increase in FFA from 0.0503 ± 0.000041% as oleic acid up to 0.0602 ± 0.000006% as oleic acid after 180 days storage at ambient temperature. At 50°C, control reached the maximum value upto 0.0646 ± 0.00005 as oleic acid after 90. From figure 6 and figure 7 it can be seen that sunflower oil treated with OF001 @ 0.02% and OF002 @ 0.02% showed lesser increase in FFA value of sunflower oil as compared to other treatments.

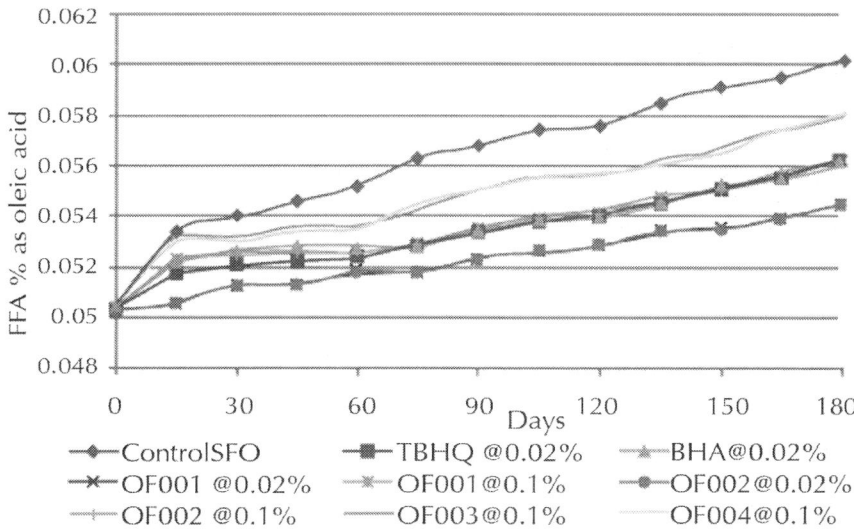

Figure 6: Effect of antioxidant type and concentration on free fatty acid value of sunflower oil at ambient temperature.

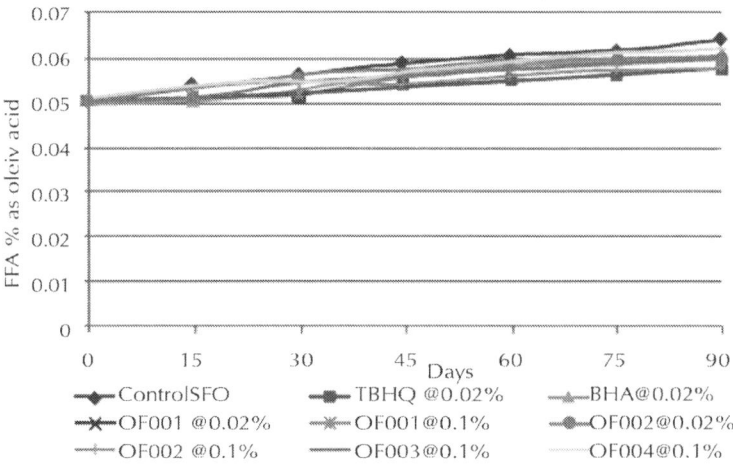

Figure 7: Effect of antioxidant type and concentration on free fatty acid value of sunflower oil at 50°C.

Effect on Total Polar Compounds

The purpose of this method is to determine the level of polar, oxidized components in a sample. Faster rate of formation of polar components is cumulative indication of degradation of oil. As shown in figure 8, control sunflower oil reached the maximum value of total polar compounds (TPC) upto 5.931% from initial TPC value of 2.734% after 180 days storage at ambient temperature. Sunflower oil treated with OF001 @ 0.1% showed the minimum value (4.04%) for TPC after 180 days storage at ambient temperature whereas TBHQ gave an increase up to 4.319%. From figure 8 and figure 9, it is seen that addition of antioxidant effectively reduced the formation of polar compounds at ambient temperature and 50°C. Control sunflower oil reached maximum TPC value of 8.1434% after 90 days storage at 50°C. Sunflower oil treated with OF001 @ 0.1% was found to be more effective than TBHQ at ambient temperature whereas at 50°C OF001 @ 0.1% reached the TPC value up to 5.4325% against TBHQ (5.1277%).

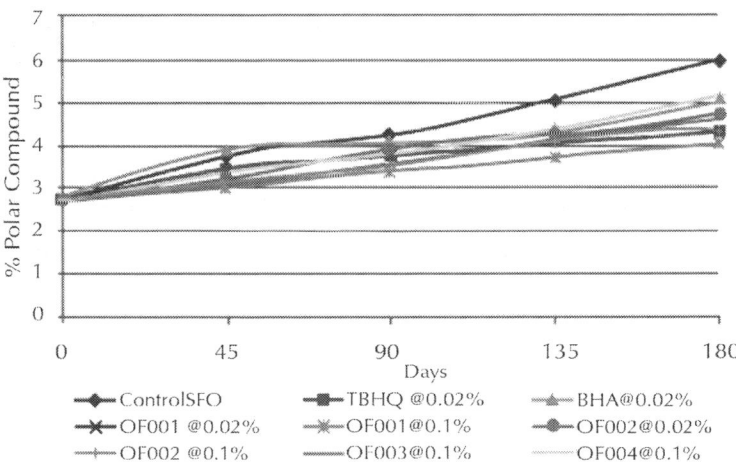

Figure 8: Effect of antioxidant type and concentration on total polar compounds of sunflower oil at ambient temperature.

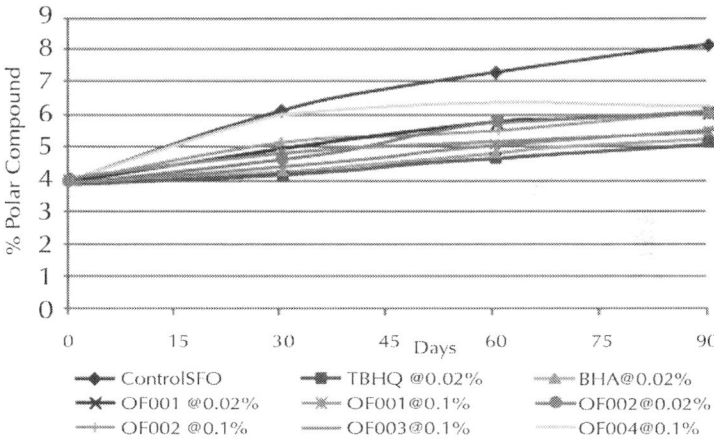

Figure 9: Effect of antioxidant type and concentration on total polar compounds of sunflower oil at 50°C.

CONCLUSIONS

During storage of sunflower oil, OF001 and OF002 effectively retarded the primary oxidation up to 120 days while OF003 could prevent the primary oxidation up to 90 days at ambient temperature. At 50°C, all treatments including natural antioxidants and synthetic antioxidant, incorporation of TBHQ were found suitable to retard primary oxidation up to 30 days while oil treated with BHA and control sunflower oil could not prevent primary oxidation even for 15 days. OF001, OF002, OF003 and OF004 were found to inhibit the p-anisidine value of sunflower oil during both storage conditions as effectively as synthetic antioxidants viz. TBHQ and BHA. All the treatments were found to retard secondary oxidation products as compared to control sunflower oil. OF001 and OF002 @ 0.02% showed least increase in FFA value of sunflower oil among all treatments including sunflower oil treated with TBHQ and BHA during both storage conditions. During storage, total polar compound were not found to increase significantly. All the treatments, both natural and synthetic antioxidants were found to keep the value of total polar compound lesser than the control sample. Sunflower oil exhibited more susceptibility to oxidative deterioration as peroxide value of control sunflower was found to increase rapidly after 15 days of storage at ambient temperature. For sunflower oil, OF002 @ 0.1% was found effective to restrict the peroxide value within threshold limit up to 120 days of storage.

ACKNOWLEDGEMENTS

Plant extract samples provided by M/s Synthite Industries Ltd., Cochin, Kerala, India for experimental work is thankfully acknowledged.

REFERENCES

1.   Gunstone F D (2004) Vegetable Oils in Food Technology: Composition, Properties and Uses. (2nd Ed.). USA.

2.   Yanishlieva N, Pokorny J, Gordon M (2001) Antioxidants in Food: Practical Applications. (1st edn), Cambridge, USA.

3. Bruni R, Muzzoli M, Ballero M, Loi MC, Fantin G, et al. (2004) Tocopherols, fatty acids and sterols in seeds of four Sardinian wild Euphorbia species. Fitoterapia 75: 50-61.

4. Frutos MJ, Hernandez-Herrero JA (2005) Effects of rosemary extract (Rosmarinus officinalis) on the stability of bread with an oil, garlic and parsley dressing. Food Science and Technology-LWT 38: 651–655.

5. Hras AR, Hadolin M, Knez Z, Bauman D (2000) Comparison of antioxidative and synergistic effects of rosemary extract with a-tocopherol, ascorbyl palmitate and citric acid in sunflower oil. Food Chem 71: 229-233.

6. Williams RJ, Spencer JPE, Rice-Evans C (2004) Flavonoids: Antioxidants or signalling molecules? Free Radical Biology and Medicine 36: 838–849.

7. Amarowicz R, Naczk M, Shahidi F (2000) Antioxidant activity of various fractions of non-tannin phenolics of canola hulls. J Agri Food Chem 48: 2755–2759.

8. Aruoma OI, Halliwell B, Aeschbach R, Loligers J (1992) Antioxidant and pro-oxidant properties of active rosemary constituents: Carnosol and carnosic acid. Xenobiotica 22: 257–268.

9. Sherwin ER (1990) Antioxidants. In A. L. Branen, P. M. David- son & S. Salminen, Food antioxidants. New York: Marcel Dekker Inc.

10. Wanasundara UN, Shahidi F (1998) Antioxidant and pro-oxidant activity of green tea extracts in marine oils. Food Chem 63: 335-342.

11. Chu YH, Hsu HF (1999) Effects of antioxidants on peanut oil stability. Food Chem 66: 29-34.

12. Peter KV (2001) Handbook of Herbs and Spices. Vol. 1. (1st ed.). Cambridge, London: Woodhead Publishing Limited, (Chapter 1).

13. Okamura N, Fujimoto Y, Kuwabara S, Yagi A (1994) High performance liquid chromatographic determination of carnosic acid and carnosol in Rosmarinus officinalis and Salvia officinalis. J Chromatogr A 679: 381–386.

14. SingletonVL, Orthofer R, Lamuela-Raventos RM (1999) Analysis of total phenols and 260other oxidation substrates and antioxidants

by means of Folin–Ciocalteu reagent. Methods in Enzymology 299: 152–178.

Oxidative DNA Damage Preventive Activity and Antioxidant Potential of Plants Used in Unani System of Medicine

Mehar Darukhshan Kalim, Dipto Bhattacharyya, Anindita Banerjee, and Sharmila Chattopadhyay

Plant Biotechnology Laboratory, Drug Development/Diagnostics & Biotechnology Division, Indian Institute of Chemical Biology (A unit of Council of Scientific & Industrial Research), 4, Raja S.C. Mullick Road, Kolkata 700 032, India

ABSTRACT

Background

There is increasing recognition that many of today's diseases are due to the "oxidative stress" that results from an imbalance between the formation and neutralization of reactive molecules such as reactive oxygen species (ROS) and reactive nitrogen species (RNS), which can be removed with antioxidants. The main objective of the present study was to evaluate the antioxidant activity of plants routinely used in the Unani system of medicine. Several plants were screened for radical scavenging activity, and the ten that showed promising results were selected for further evaluation.

Methods

Methanol (50%) extracts were prepared from ten Unani plants, namely *Cleome icosandra, Rosa damascena, Cyperus scariosus, Gardenia gummifera, Abies pindrow, Valeriana wallichii, Holarrhena antidysenterica, Anacyclus pyrethrum, Asphodelus tenuifolius* and *Cyperus scariosus*, and were used to determine their total phenolic, flavonoid and ascorbic acid contents, in vitro scavenging of DPPH·, ABTS·+, NO, ·OH, O_2·⁻ and ONOO⁻, and capacity to prevent oxidative DNA damage. Cytotoxic activity was also determined against the U937 cell line.

Results

IC_{50} values for scavenging DPPH·, ABTS·+, NO, ·OH, O_2·⁻ and ONOO⁻ were in the ranges 0.007 ± 0.0001 - 2.006 ± 0.002 mg/ml, 2.54 ± 0.04 - 156.94 ± 5.28 µg/ml, 152.23 ± 3.51 - 286.59 ± 3.89 µg/ml, 18.23 ± 0.03 - 50.13 ± 0.04 µg/ml, 28.85 ± 0.23 - 537.87 ± 93 µg/ml and 0.532 ± 0.015 - 3.39 ± 0.032 mg/ml, respectively. The total phenolic, flavonoid and ascorbic acid contents were in the ranges 62.89 ± 0.43 - 166.13 ± 0.56 mg gallic acid equivalent (GAE)/g extract, 38.89 ± 0.52 - 172.23 ± 0.08 mg quercetin equivalent (QEE)/g extract and 0.14 ± 0.09 - 0.98 ± 0.21 mg AA/g extract. The activities of the different plant extracts against oxidative DNA damage were in the range 0.13-1.60 µg/ml. Of the ten selected plant extracts studied here, seven - *C. icosandra, R. damascena, C. scariosus, G. gummifera, A. pindrow, V. wallichii* and *H. antidysenterica* - showed moderate antioxidant activity. Finally, potentially significant oxidative DNA damage preventive activity and antioxidant activity were noted in three plant extracts: *C. icosandra, R. damascena* and *C. scariosus*. These three plant extracts showed no cytotoxic activity against U937 cells.

Conclusions

The 50% methanolic extracts obtained from different plant parts contained significant amounts of polyphenols with superior antioxidant activity as evidenced by the scavenging of DPPH·, ABTS·+, NO, ·OH, O_2·− and ONOO−. *C. icosandra, R. damascena* and *C. scariosus* showed significant potential for preventing oxidative DNA damage and radical scavenging activity, and the *G. gummifera, A. pindrow, V. wallichii, H. antidysenterica, A. pyrethrum, A. tenuifolius* and *O. mascula* extracts showed moderate activity. The extracts of *C. icosandra, R. damascena* and *C. scariosus* showed no cytotoxicity against U937 cells. In conclusion, these routinely used Unani plants, especially *C. icosandra, R. damascena* and *C. scariosus*, which are reported to have significant activity against several human ailments, could be exploited as potential sources of natural antioxidants for plant-based pharmaceutical industries.

BACKGROUND

The World Health Organization estimates that 80% of the world's inhabitants rely mainly on traditional medicines for their health care [1]. Herbs contain some of the most powerful natural antioxidants and are highly prized for their antioxidant and anti-ageing effects.

Natural products offer an untold diversity of chemical structures. These compounds often serve as lead molecules, the activities of which can be enhanced by chemical manipulation and by de novo synthesis [2,3]. To date, many medicinal plants have proved successful in combating various ailments, leading to mass screening for their therapeutic components.

Antioxidants are widely used as ingredients in dietary supplements and are exploited to maintain health and prevent oxidative stress-mediated diseases such as cancer, atherosclerosis, diabetes, inflammation and ageing. Recently, many antioxidants have been isolated from different plant materials [4-6]. Natural antioxidants are also in high demand for application as nutraceuticals and as food additives because of consumer preferences [7,8]. In addition to their uses in medicine, these compounds are used in industry e.g. as preservatives in food and cosmetics and for preventing the degradation of rubber and gasoline. Antioxidants are also used as additives to help guard against food deterioration. Among natural antioxidants, plant polyphenols+ are especially important [9]. Today, the search for natural compounds rich in antioxidant, anticancer and antimicrobial properties is escalating because of their importance in controlling many chronic disorders such as cancer and cardiovascular diseases [5]. It has been estimated that approximately two-thirds of anticancer drugs approved worldwide up to 1994 were derived from plant sources [10].

It is increasingly being realized that many of today's diseases are due to the "oxidative stress" that results from an imbalance between the formation and neutralization of prooxidants [6]. These excess free radicals react with biological macromolecules such as proteins, lipids and DNA in healthy human cells and this results in the induction of carcinogenesis, atherosclerosis, cardiovascular diseases, ageing and inflammatory diseases [11,12]. These harmful radicals have to be eliminated from biological systems by enzymes such as superoxide dismutase, catalase and peroxidase, or compounds such as ascorbic acid, tocopherol and glutathione, which possess antioxidant properties.

Unani medicine, a form of traditional medicine widely practiced in India and the rest of the Indian subcontinent, is orientated towards prevention, health maintenance and treatment. Herbal products are regularly used in traditional medicines such as Ayurveda and Unani, which strengthen body defences [13]. Unani therapies cure the diseases without such side effects even after they have been consumed for a long time with a wide spectrum of therapeutic activity. Unani therapies are known to be relatively economic and are most popular amongst people because they are safe and have time-tested efficacy. They contain vitamins, minerals, active steroids, alkaloids, glycosides and tannins as well as a variety of antioxidants in a biologically natural state.

In this study we screened several Unani plants regularly prescribed by local practitioners to cure various ailments. Some of them have also been reported to have antioxidant activity [14-16]. These plants have been reported to show several activities, e.g. the tubers of *C. scariosus* are credited with astringent, diaphoretic, diuretic etc. properties; *R. damascena* flower buds are astringent and are used in cardiac troubles etc.; *V. wallichii* roots possess antiplasmodic properties and have a depressant effect on the central nervous system; and finely powdered *O. mascula* roots boiled with milk form a nutritious item of diet that is administered for diabetes, phthisis, chronic diarrhoea and dysentery [17]. Fresh leaf juice from the plant *C. icosandra* has been taken orally for toothache, whereas the seeds have been claimed to have anthelmintic properties. The bark of *H. anidysenterica* has astringent, antidysenteric, anthelmintic, stomachic, febrifugal etc. properties [18]. Gum obtained from *G. gummifera* is used internally in dyspepsia accompanied by flatulence and is also considered antispasmodic and carminative, antiseptic and stimulating. The roots of *A. pyrethrum* excite a remarkable flow of saliva and possess stimulating and rubefacient properties, whereas the leaves are used as carminatives, astringents etc. and the seeds are considered to be diuretic. *A. pindrow* leaves are carminative, astringent and antipyretic, and also used in asthma and bronchitis. *A. tenuifolius* seeds are considered to be diuretic [19]. These plants are edible and so considered safe [20]. The plants and their parts evaluated in this study are listed in Table 1. To date, few studies have been carried out to evaluate their antioxidant properties. Here we report our evaluation of their in vitro antioxidant potential, including the scavenging of DPPH, ABTS^{+}, NO, OH, O$_2^{-}$ and ONOO^{-}, along with their activity in preventing oxidative DNA damage and cytotoxicity against U937 cells. To evaluate the mechanism of action of anti-oxidant properties of these plants, the total polyphenol and total flavonoid contents of all ten extracts were determined.

Table 1: Properties of Unani plants used in this study

Family	Common name [1,2]	Botanical name	Parts used	Uses	Class of compound	Name of compound
Capparidaceae	Dog mustard/Hurhur	*Cleome icosandra*	Seed[3], leaves, flower	It is vata and kapha suppressant, a good pain reliever, also a good antibacterial and antiwormal, reduces pus formation in the wounds, helpful in convulsions, has a good effect on digestive tract and improves indigestion condition in the body, increases sweating in the body	Coumarino-lignans	Cleomiscosin A, C,
Rosaceae	Golap/Gulab	*Rosa damascena*	Petals[3]	Gulkhand made by the mixture of rose petals and white sugar in equal proportion act as the tonic and laxative, used as herbal tea in the treatment of cold and cough	Components of Essential oil; Flavonol and their glycosides; Flavonoid	Citronellol, Geraniol, Linalool etc. Glycosides of Kaempferol and Quercetin; Quercetin, Kaempferol
Cyperaceae	Umbrella's edge/Nagarmotha	*Cyperus scariosus*	Root[3]	Intestinal disorders, astringent, diaphoretic, diuretic, desiccant, cordial, and stomachic properties, treatment of gonorrhea	Essential Oils	Some volatile compounds reported till date from the oil are atchoulanol, selina-4, etc.
Rubiaceae	Gummy gardenia/Dikamali	*Gardenia gummifera*	Resin[3]	Kapha skin disease, indigestion, worm infestation, diarrhea and infections, the resin has antiseptic property	Flavonoids, Flavone; Seco-cycloartenol derivatives	Gardenin E; Dikmaliartanes
Pinaceae	Himalayan fir/Dodimma	*Abies pindrow*	Bark[3], leaves, trunk	Disorders with inflammatory system	Proanthocyanidins	Potential rich sources up to 5% of bark weight

Family	English name[1]/Hindi name[2]		Plant part used[3]	Uses		
Valerianaceae	Gilgiti valerian/Ganeshpawrobati	Valeriana wallichii	Roots[3]	Antisplasmodic, stimulant, calmative and stomachic, useful in diseases of eye and liver, used as a remedy for hysteria, hypochondriasis, nervous unrest and emotional arrest, also useful in clearing voice and acts as a stimulant in advance stage of fever and nervous disorder	Essential oil and volatile oil / Iridous	0.3-1% volatile oil content / Valtrate, didrovaltrate, acetovaltate, etc.
Apocynaceae	Tellycherry bark/Kurchi	Holarrhena antidysenterica	Bark[3]	The bark is used as an astringent, anthelmintic, antidontalgic, stomachic, febrifuge, antidropsical, diuretic, in piles, colic, dyspepsia, chest affections and as a remedy in diseases of the skin and spleen, use as a well-known drug for amoebic dysentery and other gastric disorders	Poly-phenolics / Plant sterols	-sitosterol
Asteraceae	Pellitory/Akarkara	Anacyclus pyrethrum	Root[3]	Stimulant, sialogogue, and rubefacient properties	Glycosides / Phenolic acids / Lignans	Flavone glycosides / Chlorogenic acid / Sesamin
Orchidaceae	Orchid/Salebpanja	Orchis mascula	Flower, tuber[1]	Tonic, aphrodisiac, yield a lot of mucilage with water and form a jelly that is supposed to be nutritious and useful in diarrhea, dysentery, and chronic fever	Bitter principle and a volatile oil	
Asphodelaceae	Onion weed/Jangli pyaz	Asphodelus tenuifolius	Bulb[1], seed	As diuretic and on inflammation	Flavonoid	Luteolin and its glycosides

[1]English name [2]Hindi name [3]Plant part used in this study

Kalim et al. BMC Complementary and Alternative Medicine 2010 10:77, doi:10.1186/1472-6882-10-77

METHODS

Plant Materials and Extraction Procedure

Plants were collected from, and authenticated by, a Unani medical practitioner in Kolkata, India who regularly prescribes these materials. The different plant parts were shade-dried at room temperature (25°C) with occasional turning of the plants upside down for 5-7 days, and then ground to coarse powder with a mechanical grinder. The powdered plant materials (2 g) were extracted with 50 ml of aqueous methanol (50:50) for three consecutive days with intermittent stirring (1 h stirring at every 12 h interval) using magnetic stirrer until the extracts were light colored. The combined extracts were filtered and evaporated under reduced pressure in a rotary vacuum evaporator (Eyela NVC-2100--Rotary Evaporator, water bath temperature maintained at 40°C and 356 mm Hg, Eyela NCB-1200 Chiller unit temperature maintained at 7.5°C). The aqueous layer was lyophilized (at -45°C) and the dry powder was stored at -20°C for future use.

Reagents Used

2, 2-Diphenyl-1-picrylhydrazyl (DPPH), thiobarbituric acid (TBA), Folin--Ciocalteu's phenol reagent, butylated hydroxytoluene, agarose and ethidium bromide were purchased from Sigma-Aldrich, USA. 2, 2'-Azinobis-(3-ethyl-benzothiazoline-6-sulfonic acid) (ABTS), potassium persulphate, aluminium chloride, iron (III) chloride, and iron (II) sulphate were obtained from MP Biomedicals, USA. 2-Deoxy-D-ribose and ascorbic acid were procured from Himedia Laboratories Pvt. Ltd., Mumbai, India. The QIAprep Spin Miniprep Kit was purchased from Qiagen, Germany. All other chemicals and reagents used were of analytical grade.

Analytical Studies

For all the analytical studies, absorbance was measured using a Shimadzu UV-Visible Pharmaspec 1700 spectrophotometer.

Determination of Total Phenolic Content (TPC)

The total phenolic contents of the 50% methanolic plant extracts were determined with gallic acid as a positive standard [21]. Aliquots of test samples (100 µl) were mixed with 2 ml 2% Na_2CO_3 and incubated at 25°C for 2 min. After incubation, 1:1 (v/v) Folin-Ciocalteu's phenol reagent was added and the contents were mixed vigorously. The mixture was allowed to stand at 25°C for 30 min and the absorbance was measured at 720 nm. The same procedure was repeated with all standard gallic acid solutions and a standard curve was obtained. The total polyphenolic contents of the extracts were expressed in terms of gallic acid equivalents (GAE) of the plant sample.

Determination of Total Flavonoid Content (TFC)

The total flavonoid content was determined using quercetin as a positive standard and expressed in terms of quercetin equivalents (QEE) in mg/g plant sample [22]. $NaNO_2$ (150 µl, 5% w/v) was added to tubes containing plant extracts in 2.5 ml distilled water. The contents were mixed thoroughly and allowed to stand for 5 min at ambient temperature, then 1.5 ml of 10% (w/v) $AlCl_3$ were added and the mixture was allowed to stand for another 6 min. The solution was immediately mixed after addition of 1 ml 1 M NaOH. After 10 min, the absorbance was measured at 510 nm.

Determination of Total Ascorbic Acid (ASC)

ASC of plant extracts were determined according to Roe and Kuether [23] after brief modification. Blanks, standards and samples were prepared in triplicate to measure ASC. Ascorbic acid (AA) standards (0-10 mM) or samples were precipitated with 10% trichloroacetic acid followed by centrifugation. In 500 µL of supernatant, 100 µL of DTC reagent (2,4-dinitrophenylhydrazine 3%, thiourea 0.4%, and copper sulfate 0.05%) prepared in 9N sulfuric acid, was mixed and incubated at 37°C for 3 h. After the addition of 750 µL of 65% (v/v) sulfuric acid, the absorbance was recorded at 520 nm. A standard curve was prepared with AA standards, and ASC was expressed as mg AA/g of plant sample.

Determination of Free Radical Scavenging Activity

The radical scavenging activities of the plant extracts in the range 0-200 µg/ml were evaluated using DPPH·. Stock solutions of plant extracts were prepared at a concentration of 10 mg/ml and a freshly-prepared DPPH solution (100 mM) was used as described previously [7].

Scavenging of ABTS·+ was assayed to assess the antioxidant capacities of the 50% methanolic plant extracts according to the method of Re et al. [24]. The ABTS stock solution was prepared by reacting ABTS (7 mM) and potassium persulphate (2.45 mM) and allowing the mixture to stand for at least 16 h to generate ABTS·+ free radicals. The working solution was prepared by diluting the stock solution with methanol such that its absorbance reached 0.7 ± 0.02 at 734 nm ($A_{Control}$). For the assays, 1 ml of ABTS·+ working solution was mixed with 10 µl extracts of different concentrations (0-100 µg/ml). Their absorbance (A_{Sample}) was noted at 734 nm exactly 6 min after the reaction mixture was prepared. In both assays, quercetin was used as positive control. The control reaction contained no test sample. The percentage radical scavenging activity (% RSC) was calculated using the formula:

$$\% \text{ RSC} = [(A_{Control} - A_{Sample})/A_{Control}] \times 100\%$$

Determination of Hydroxyl Radical Scavenging Activity

The ·OH scavenging assay was performed as standardized before [8]. The reaction mixture consisted of different concentrations (0-100 µg/ml) of plant extract, 3.6 mM deoxyribose, 0.1 mM EDTA, 0.1 mM L-ascorbic acid, 1 mM H_2O_2 and 0.1 mM $FeCl_3.6H_2O$, and the volume was made up to 500 µl with 25 mM phosphate buffer, pH 7.4. This mixture was incubated for 1 h at 37°C, 500 µl of 1% TBA and 500 µl of 1% TCA were added, and the mixture was heated in a boiling water-bath for 15 min and then cooled. The absorbance was measured at 532 nm. The control reaction contained no test sample, and quercetin (20 μg/ml) was used as a standard. Percentage RSC was calculated as described above.

Determination of Peroxynitrite Scavenging Activity

Peroxynitrite was synthesized by the method of Beckman et al. 1994 [25]. Briefly, an acidic solution of 0.7 M H_2O_2 was mixed with an equal volume of 0.6 M potassium nitrite in an ice bath and an equal volume of ice cold 1.2 M NaOH was added. Granular MnO_2 prewashed with 1.2 M NaOH was used to remove excess H_2O_2 and the reaction mixture was left at -20°C. The concentration of peroxynitrite generated was measured spectrophotometrically at 302 nm ($\varepsilon = 1670$ M^{-1} cm^{-1}).

Peroxynitrite scavenging activity was measured according to Hazra et al. 2010 [26]. The reaction mixture consisted of 0.1 mM DTPA, 90 mM NaCl, 5 mM KCl, 12.5 μM Evans Blue, plant extracts at various doses ranging from 0-300 μg/ml, and 1 mM peroxynitrite adjusted to a final volume of 1 ml with 50 mM phosphate buffer (pH 7.4). The reaction mixture was incubated at 25°C for 30 min and the absorbance was measured at 611 nm. The percentage peroxynitrite scavenging activity was calculated by comparing the results of the test and blank samples; gallic acid served as the reference compound. All tests were conducted six times. The IC_{50} values of the extracts were calculated by regression analysis.

Determination of Non-enzymatic Superoxide Radical Scavenging Activity

Superoxide radical was generated *in vitro* by a non-enzymatic method involving the nicotinamide adenine dinucleotide-nitro blue tetrazolium-phenazine methosulphate (NADH-NBT-PMS) system following the procedure of Nishikimi et al. [27]. NBT (150 μM in 0.02 M Tris buffer, pH 8.0) was added to 1 ml of NADH solution (50 μM of NADH in 0.02 M Tris buffer, pH 8.0) in the presence of various concentrations (0-50 μg/ml) of extracts. The reactions were initiated by adding PMS (15 μM) and the absorbance was at 560 nm was measured exactly 1 min later. Results were recorded as percentage inhibition. Quercetin at various concentrations was used as standard. All tests were performed six times.

Nitric Oxide Scavenging Activity: Concentration Dependence

The scavenging activity against nitric oxide was assayed by the method of Marcocci et al. [28]. Sodium nitroprusside (0.5 ml, 5 mM in 20 mM phosphate buffer, pH 7.4, previously bubbled with argon) was added to tubes containing 0.5 ml of different plant extracts of various concentrations (0-300 µg/ml) and incubated at 25°C for 150 min. At the end of the incubation, 1 ml of Griess reagent (equal volumes of 2% w/v sulphanilamide in 5% phosphoric acid and 0.2% w/v naphthylethylenediamine dihydrochloride) was added to each sample and the absorbance was measured at 546 nm against control samples (extracts incubated with only 20 mM phosphate buffer, pH 7.4) and referred to the absorbance of standard solutions of sodium nitrite treated in the same way with Griess reagent. Results were recorded as percentage nitrite formed. Quercetin at various concentrations was used as standard.

Prevention of Oxidative DNA Damage

This was determined as described previously [8]. Plasmid DNA was isolated using a QIAprep Spin Miniprep Kit according to the manufacturer's instructions. Plasmid pBluescript II SK (-) (250 ng) was treated with $FeSO_4$, H_2O_2 and phosphate buffer (pH 7.4) in final concentrations of 0.5 mM, 25 mM and 50 mM, respectively, and test extracts at different concentrations (0-2 µg/ml). The total reaction volume was set to 12 µl and the mixture was incubated at 37°C for 1 h. After the incubation, the extent of DNA damage and the preventive effect of the test samples were analyzed on 1% agarose gels at 70 V at room temperature. Quercetin (1 mM) was used as positive control.

Gels were scanned on a Gel documentation system (GelDoc-XR, Bio-Rad, Hercules, CA, USA). Bands were quantified using discovery series Quantity One 1-D analysis software (Bio-Rad).

In Vitro Cytotoxicity Activity (MTT Assay)

The cytotoxicity of the plant extracts against the U937 cells was determined using the MTT (thiazolyl blue tetrazolium bromide) assay adapted from Kim et al. [29]. Cells were seeded into 96-well plates at 5,000-10,000 cells/well and treated with different concentrations of the plant extracts. After 48 h, MTT was added to each well and the formazan crystals were dissolved in DMSO. The absorbance was measured at 570 nm using a microplate ELISA reader. All experiments were performed in eight replicates. Percentage cell survival was calculated using the following formula:

$$\% \text{ cell survival} = [(A_t - A_b)/(A_c - A_b)] \times 100$$

Where A_t = absorbance of test sample; A_b = absorbance of blank (medium); A_c = absorbance of control (cells).

Statistical Analysis

All data were expressed as mean ± SD. Statistical analyses were performed using Microsoft Excel. The IC_{50} values were calculated by regression analysis. Values with $p < 0.05$ were considered statistically significant. The IC50 values were compared by paired t test (two-sided).

RESULTS

Total Phenolic, Flavonoid and Ascorbic Acid Contents

Ten selected plants regularly prescribed in Unini system of medicine was investigated here (Table1). 50% methanolic extracts of different parts of the were determined TPC, TFC and ASC were expressed as mg GAE/g extract, mg QEE/g extract and mg AA/g extract, respectively (Table 2).

Table 2: Total phenolic, flavonoid and ascorbic acid contents of plant extracts

Plant name	Total phenolic content mg GAE/g plant extract[1]	Total flavonoid content mg QEE/g plant extract[1]	Total ascorbic acid content mg AA/g plant extract[1]
Cleome icosandra	166.13 ± 0.56	172.23 ± 0.08	0.98 ± 0.218
Rosa damascena	142.23 ± 0.09	151.32 ± 0.51	0.82 ± 0.092
Cyperus scariosus	128.83 ± 0.32	118.93 ± 0.23	0.39 ± 0.017
Gardenia gummifera	82.72 ± 0.03	87.32 ± 0.13	0.49 ± 0.029
Abies pindrow	76.82 ± 0.13	63.82 ± 10.71	0.47 ± 0.079
Valeriana wallichii	72.13 ± 0.51	74.32 ± 0.21	0.55 ± 1.012
Holarrhena antidysenterica	69.12 ± 0.35	60.42 ± 0.34	0.42 ± 0.077
Anacyclus pyrethrum	62.89 ± 0.43	38.89 ± 0.52	0.37 ± 0.12
Orchis mascula	12.52 ± 0.57	12.11 ± 1.20	0.33 ± 0.073
Asphodelus tenuiofolius	15.74 ± 0.98	11.98 ± 0.74	0.14 ± 0.091

Results are mean ± SD from three sets of independent experiments, each set in triplicate

Kalim et al. BMC Complementary and Alternative Medicine 2010 10:77, doi: 10.1186/1472-6882-10-77

The mean values of total phenols ranged from 62.89 ± 0.43 to 166.13 ± 0.56 mg GAE/g, flavonoids from 38.89 ± 0.52 to 172.23 ± 0.08 mg QEE/g extract and ASC from 0.14 ± 0.091 to 0.98 ± 0.218 AA/g extract. The highest TPC was observed in *C. icosandra* (166.13 ± 0.56 mg GAE/g extract), followed by *R. damascena* (142.23 ± 0.09 mg GAE/g extract) and *C. scariosus*(128.83 ± 0.32 mg GAE/g extract). For TFC, *C. icosandra* (172.23 ± 0.08 mg QEE/g extract) showed the highest content, also followed by *R. damascena* (151.32 ± 0.51 mg QEE/g extract) and *C. scariosus* (118.93 ± 0.23 mg QEE/g extract). The ASC contents were 0.98 ± 0.21, 0.82 ± 0.092 and 0.39 ± 0.017 mg AA/g extract in *C. icosandra, R. damascena* and *C. scariosus*, respectively followed by other extracts.

DPPH˙ scavenging Activity

The free radical scavenging activities of the extracts, as measured by the ability to scavenge DPPH free radicals, were compared with quercetin; the lower the IC_{50}, the stronger the scavenging activity (Table 3). The maximum % inhibition of the following extracts was noted like 85% in *C. icosandra* at 12.37 ± 1.09 µg/ml, 83% in *R. damascena* at 17.19 ± 0.23 µg/ml and 80.52% in *C. scariosus* at 17.71 ± 0.71 µg/ml followed by other extracts.

Table 3: IC50 values of plant extracts (µg/ml)

Plant Name	DPPH·	ABTS·+	·OH	NO	O_2·-	ONOO-
Cleome icosandra	7.28 ± 0.37**	2.54 ± 0.04***	20.13 ± 0.01***	152.23 ± 3.51***	30.96 ± 0.98***	532.85 ± 15.93*
Rosa damascena	10.36 ± 0.02***	3.57 ± 0.11**	23.01 ± 0.03**	273.18 ± 3.52***	42.10 ± 0.82[NS]	637.57 ± 52.93**
Cyperus scariosus	11.10 ± 0.37***	6.27 ± 0.44**	18.23 ± 0.038***	240.31 ± 4.28***	28.85 ± 0.23***	590.23 ± 2.37*
Gardenia gummifera	82.33 ± 0.31***	11.62 ± 0.21**	34.33 ± 0.07***	--	45.39 ± 0.87***	890.32 ± 52.23***
Abies pindrow	84.23 ± 1.50**	19.10 ± 0.21***	31.43 ± 0.07***	286.59 ± 3.89***	74.54 ± 9.28***	987.42 ± 17.4***
Valeriana wallichii	86.61 ± 0.89**	21.26 ± 0.18***	37.92 ± 0.07***	--	78.35 ± 0.57***	943.12 ± 27.82***
Holarrhena antidysenterica	98.84 ± 0.31***	29.92 ± 0.25***	29.23 ± 0.01**	211.34 ± 2.12***	83.49 ± 0.59***	880.51 ± 9.99***
Anacyclus pyrethrum	467.10 ± 0.27***	31.76 ± 0.27***	41.22 ± 0.04***	--	83.49 ± 0.59***	1.137 ± 0.003[1]***
Orchis mascula	1.098 ± 0.009[1]***	--	47.82 ± 0.20**	--	537.87 ± 93.12**	3.114 ± 0.09[1]***
Asphodelus tenuifolius	2.006 ± 0.002[1]***	156.94 ± 5.28***	50.13 ± 0.04***	--	425.92 ± 78.12***	3.390 ± 0.031[1]***
Reference Compound						
Quercetin	3.21 ± 0.11	1.34 ± 0.08	7.42 ± 0.32	18.23 ± 0.42	41.98 ± 0.95	
Gallic acid				--		820.12 ± 27.34

Results are mean ± SD (n = 3), each set in triplicate

Units of IC50: (mg/ml)[1]

* p < 0.05; ** p < 0.01;*** p < 0.001; NS = Non significant

Kalim et al. BMC Complementary and Alternative Medicine 2010 10:77, doi:10.1186/1472-6882-10-77

ABTS·+ scavenging Activity

The IC_{50} values of the plant extracts were also determined for ABTS·+ (Table 3). Significant activity was noted with *C. icosandra, R. damascena* and *C. scariosus* with 98.23% inhibition at 6.98 ± 0.07 µg/ml, 91.83% inhibition at 8.42 ± 0.13 µg/ml and 72% inhibition at 8.32 ± 0.09 µg/ ml, respectively.

·OH scavenging Activity

The ·OH scavenging potentials manifested by the different plant extracts were also evaluated by decreased formation of the chromogen in the Fenton reaction. The ·OH scavenging activities of the 50% methanolic extracts correlated with protection against DNA damage, as shown in Table 3. Best scavenging activity was noted with *C. icosandra, R. damascena* and *C. scariosus* which showed 67.08% inhibition at 34.54 ± 0.92 µg/ml, 69.7% inhibition at 23.48 ± 0.85 µg/ml and 67.2% inhibition at 29.33 ± 0.43 µg/ml, respectively.

Peroxynitrite scavenging Activity

In all the extracts tested, peroxynitrite-scavenging activity was concentration dependent. The scavenging activities was, 72% inhibition at 766.08 ± 12.23 µg/ml, 78% inhibition at 993.72 ± 52.34 µg/ml and 69% inhibition at 814.2 ± 37.89 µg/ml for *C. icosandra, R. damascena and C. scariosus* respectively. Hence these three extracts have superior activity than that of gallic acid standard. Howerver, IC_{50} values of *H. antidysenterica* (880 ± 9.99 µg/ml) and *G. gummifera* (890 ± 52.23 µg/ml) were comparable to the standard (i.e. 820.12 ± 47.2 µg/ml) in peroxynitrite scavenging potential (Table 3).

Superoxide scavenging Activity

As is evident from Table 3, the extracts of C. *icosandra* (73% inhibition at 45.20 ± 8.25 µg/ml),R. *damascena* (81% inhibition at 68.20 ± 7.23 µg/ml) and C. *scariosus* (78.23% inhibition at 45.23 ± 0.37 µg/ml) also caused considerable scavenging of superoxide anion in comparison to the reference compound quercetin. The IC_{50} values for the superoxide scavenging activities of extracts and the reference standard are shown in Table 3. As evident from results, C. *scariosus*(28.85 ± 0.23 µg/ml) was able to quench superoxide radicals more effectively than the reference compound quercetin (41.98 ± 0.95 µg/ml).

Nitric Oxide scavenging Activity

C. *icosandra* showed significant nitric oxide scavenging activity than that of other plant extracts having 69% inhibition at 210.07 ± 18.27 µg/ml. However modest scavenging activity was also noted with R. *damascena* (73.9% at 398.84 ± 52.1 µg/ml) and C. *scariosus* (72.24% at 350.85 ± 12.3 µg/ml) respectively. IC_{50} values were presented in Table 3 along with reference compound quercetin (at 18.23 ± 0.42 µg/ml).

Prevention of Oxidative DNA Damage by Plant Extracts

To assess the prevention of oxidative DNA damage by the plant extracts further, the preventive effects were evaluated over Fenton-induced damage of pBluescript II SK (-) supercoiled DNA maintained in E. *coli* XL-1 Blue strain. Control pBS DNA showed two bands, one of open circular DNA, which was hardly visible, and one of supercoiled DNA. Combined treatment with $FeSO_4$ and H_2O_2 in the absence of plant extract led to the formation of open circular DNA by strand scission of the supercoiled DNA, whereas the plant extracts at different concentrations showing optimum activity prevented this strand scission. The maximum prevention of DNA damage was shown byC. *scariosus* at 0.13 µg/ml whereas C. *icosandra* showed the same activity at 0.16 µg/ml. R. *damascena, A. pindrow, G. gummifera, O. mascula, A. pyrethrum, A. tenuifolius, H. antidysenterica* and V. *wallichii* showed the same preventive activity at 0.2 μg/ml, 0.22 μg/ml, 0.53 μg/ml, 1.60 μg/ml, 1.52 μg/ml, 1.85 μg/ml, 0.1 μg/mi and 1 μg/ml, respectively. Among these plant extracts, *C. scariosus, C. icosandra, R. damascena* and *H. antidysenterica* provided the most effective prevention of DNA damage, as shown in Figure 1. Densitometric analysis confirmed the experimental data (Figure 2).

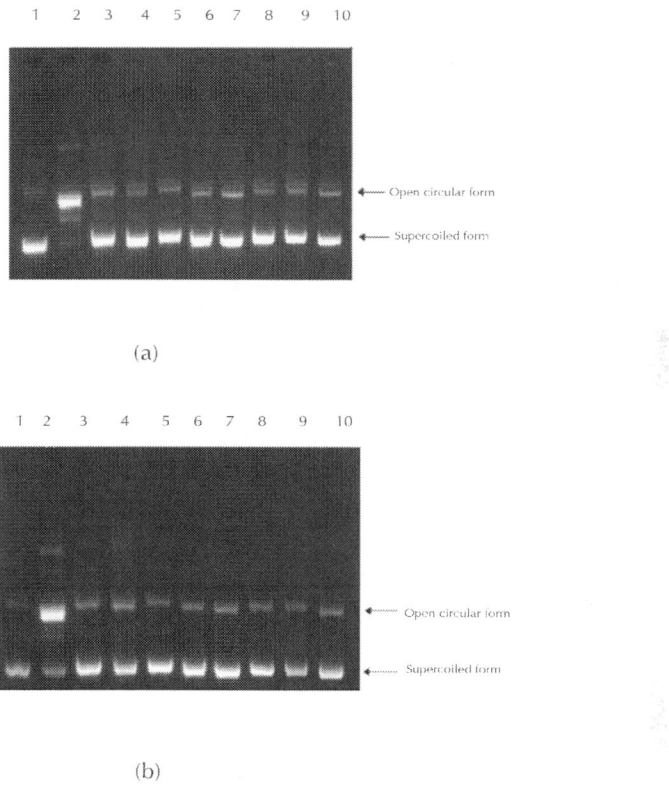

(a)

(b)

Figure 1: Electrophoresis patterns of pBluescript II SK (--) DNA breaks by ˙OH generated from the Fenton reaction and prevented by different plant extracts. (a) Lane 1: untreated control DNA (250 ng); lane 2: $FeSO_4$ (0.5 mM) + H_2O_2 (25 mM) + DNA (250 ng); lane 3: only H_2O_2 (25 mM) + DNA (250 ng); lane 4: only $FeSO_4$ (0.5 mM) + DNA (250 ng); lanes 5--10: $FeSO_4$ (0.5 mM) + H_2O_2 (25 mM) + DNA (250 ng) in the presence of quercetin (1 mM), *Gardenia gummifera*(0.53 μg/ml), *Abies pindrow* (0.29 μg/ml), *Asphodelus tenuifolius* (1.85 μg/ml), *Anacyclus pyrethrum* (1.52 μg/ml) and *Orchis mascula* (1.60 μg/ml), respectively (n = 3). (b) Lane 1: untreated control DNA (250 ng); lane

2: FeSO$_4$(0.5 mM) + H$_2$O$_2$ (25 mM) + DNA (250 ng); lane 3: only H$_2$O$_2$ (25 mM) + DNA (250 ng); lane 4: only FeSO$_4$ (0.5 mM) + DNA (250 ng); lanes 5--10: FeSO$_4$ (0.5 mM) + H$_2$O$_2$ (25 mM) + DNA (250 ng) in the presence of quercetin (1 mM), *Holarrhena antidysenterica* (0.28 µg/ml), *Valeriana wallichii* (1 µg/ml), *Rosa damascena* (0.20 µg/ml), *Cleome icosandra*(0.16 µg/ml) and *Cyperus scariosus* (0.13 µg/ml), respectively (n = 3).

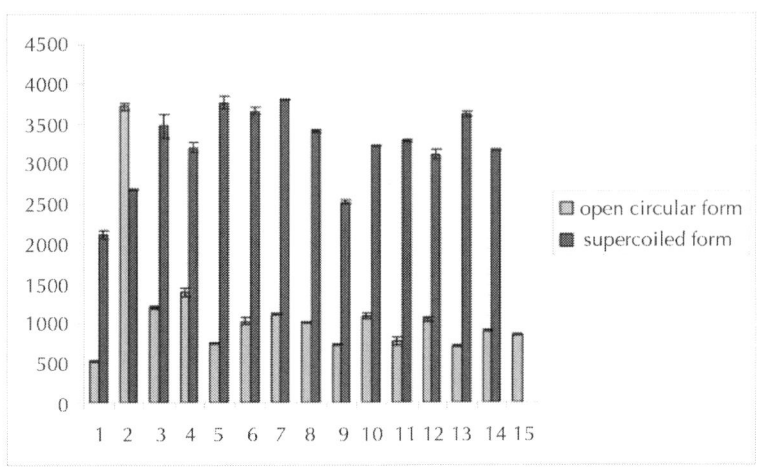

Figure 2: Densitometric analysis of open circular and supercoiled DNA damage induced by ˙OH generated from the Fenton reaction in the presence of plant extracts. Lane 1: untreated control DNA (250 ng); lane 2: FeSO$_4$ (0.5 mM) + H$_2$O$_2$ (25 mM) + DNA (250 ng); lane 3: only H$_2$O$_2$ (25 mM) + DNA (250 ng); lane 4: only FeSO$_4$ (0.5 mM) + DNA (250 ng); lanes 5--15: FeSO$_4$ (0.5 mM) + H$_2$O$_2$ (25 mM) + DNA (250 ng) in the presence of quercetin (1 mM),*Gardenia gummifera* (0.528 µg/ml), *Abies pindrow* (0.29 µg/ml), *Asphodelus tenuifolius*(1.85 µg/ml), *Anacyclus pyrethrum* (1.52 µg/ml), *Orchis mascula* (1.60 µg/ml) *Holarrhena antidysenterica* (0.28 µg/ml), *Valeriana wallichii* (1 µg/ml), *Rosa damascena* (0.2 µg/ml),*Cleome icosandra* (0.16 µg/ml) and *Cyperus scariosus* (0.128 µg/ml) respectively. Values represent mean ± SD (n = 3). The differences were considered statistically significant if $p < 0.05$.

Correlation between the TPC or TFC with the Antioxidant Activity

Correlation of phenolics content and antioxidant activity of three plant extracts with superior antioxidant activity was determined. Results showed that the correlation coefficients of total phenolics and flavonoid contents of *C. icosandra* were greater than 0.9 ($R = 0.9995$; $R = 0.9919$ respectively), the same of *R. damascena* $R = 0.9830$; $R = 0.9848$) and *C. scariosus* ($R = 0.9604$; $R = 0.9910$) was comparative as shown in Figure 3. This signified that the oxidative DNA damage preventive activity as well as antioxidant potential of these three effective plant extracts could be strictly correlated with their total phenolics and flavonoid contents.

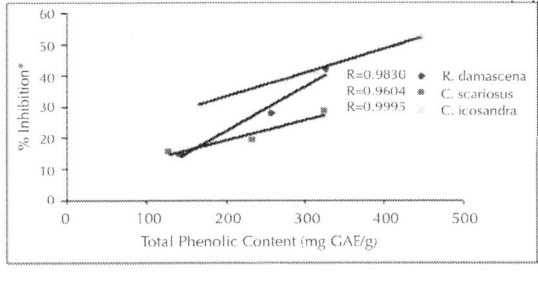

(a)

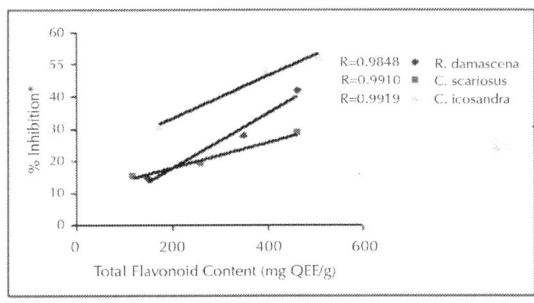

(b)

Figure 3: Correlation of % inhibition with Phenolic and Flavonoid contents. The relationship between (a) total phenolics content or (b) total flavonoid content in *R. damascena, C. scariosus, C. icosandra* and their antioxidant activity. The correlation analysis was described as linear correlation coefficient (R). The differences were considered statistically significant if $p < 0.05$. % Inhibition* = free radical scavenging activity as determined by ABTS assay.

Cytotoxicity

The cytotoxic activities of the extracts of *R. damascena, C. scariosus* and *C. icosandra* at different concentrations (0.2 and 2 µg/ml) were determined against U937 cells (Table 4). These extracts showed no cytotoxicity against U937 cells in comparison to doxorubicin.

Table 4: Cytotoxic activity of three plants at two different concentrations

Time	Rosa damascena		*Cyperus scariosus*		*Cleome icosandra*		Doxorubicin 25 ng/ml
	0.2 µg/ml	2 µg/ml	0.2 µg/ml	2 µg/ml	0.2 µg/ml	2 µg/ml	
00 h	100%	100%	100%	100%	100%	100%	100%
24 h	97.46%	95.77%	96.37%	96.47%	99.72%	96.81%	42%
48 h	97.58%	96.95%	96.20%	96.06%	99.39%	97.03%	0%
72 h	97.37%	97.08%	96.59%	96.11%	98.86%	96.75%	0%

Results are mean from three sets of experiments, each set in five replicate.

Kalim et al. BMC Complementary and Alternative Medicine 2010 10:77, doi: 10.1186/1472-6882-10-77

DISCUSSION

As a part of a concerted effort to develop herbal antioxidants from natural sources, we investigated several plants regularly prescribed in the Unani system of medicine against various human ailments. For initial free radical screening, DPPH assay followed by an ABTS assay was used which showed significant activity in *C. icosandra, R. damascena* and *C. scariosus*. To evaluate this potential more specifically, extracts were checked for ·OH scavenging and the highest activity was noted with *C. icosandra, R. damascena* and *C. scariosus*, corroborating the previous assay. Significant NO scavenging was noted with *C. icosandra,* followed sequentially by *H. antidysenterica, C. scariosus* and *R. damascena*. Peroxynitrite scavenging by *C. icosandra, C. scariosus,* and *R. damascena* was significantly greater than by the reference compound, whereas *H. antidysenterica* and *G. gummifera* showed similar activity to the standard compound. O_2^- scavenging activity was also significant in the extracts of *C. icosandra, C. scariosus* and *R. damascena*. Taken together, these findings indicate that *C. icosandra* extract is a potential candidate for free radical scavenging followed by *R. damascena* and *C. scariosus*.

Phytochemical analysis revealed significant total phenolic and flavonoid contents in the extracts of these same three plants, *C. icosandra, R. damascena* and *C. scariosus*, and these correlated with their potential radical scavenging activities. Though the ASC of these three effective extracts were insignificant, however that of *C. icosandra* and *R. damascena* was little higher than of *C. scariosus*, indicating that the antioxidant potential of *C. scariosus* arises from its total phenolic and flavonoid contents. Other plant extracts are also reported to contain polyphenolic compounds and their antioxidant activities may be related to this [30].

Flavonoids are polyphenols naturally present in nearly all plant materials [30]. Phenolic compounds are effective hydrogen donors, and this makes them good antioxidants [31]. Flavonoids are a class of compounds that remain of great scientific and therapeutic interest, and their antioxidant activity has attracted most attention. Their high antioxidant potential is attributable to their capacity to scavenge harmful ROS and other free radicals that originate from various cellular activities and lead to oxidative stress [32]. Plant-derived polyphenolic flavonoids are well known to exhibit antioxidant activity through a variety of mechanisms including scavenging of ROS, inhibiting lipid peroxidation and chelating metal ions [33]. Hence their mechanism of action is multiple; it includes the inhibition of enzymes involved in ROS generation, chelating of trace metals such as free iron and copper, and the ability to reduce highly oxidizing free radicals by hydrogen donation, thus protecting us from serious diseases such as heart attacks, strokes and even cancers. In addition, ascorbic acid acting as a chain-breaking antioxidant impairs the formation of free radicals during the biosynthesis of intracellular and extracellular substances throughout the body, including collagen, bone matrix and tooth dentine [34].

Previous studies have reported that the seeds of *C. icosandra* contain coumarino-lignans such as cleomiscosin A, B, C and D, of which A and C are reported to be antioxidants [16,35]. Collectively, these observations indicate that the free radical scavenging potential of *C. icosandra* seeds and protection they confer against oxidative

DNA damage may be attributed to their phytochemical composition. Rose essential oil is widely used in perfumery and the cosmetic industry. In addition to its perfuming effects, it is reported to possess a wide range of biochemical activities. Petals of R. damascena contain flavonol aglycons like kaempferol, quercetin and its glycosides such as kaempferol glycosides, quercitrin etc., citronellol and geraniol as the major components of its essential oil as well as tocopherol and carotene [19,36,37]. Potential antioxidant activity of rose petals may be attributed for their diversified phytochemical contents, which are consistent with earlier reports [36-38]. C. scariosus roots contain compounds such as patchoulanol, isopatchoulenone, etc. as major components of its essential oil [17]. Wei and Shibamoto [39] studied the antioxidant activities of major essential oils from several plants and reported that myristicin from parsley seeds, patchouli alcohol from patchouli, and citronellol from roses showed high antioxidant activities, which can be related to our study.

The Fenton reaction is a major physiological source of ·OH, which is produced near DNA molecules in the presence of transition metal ions such as iron and copper [40]. As previous reports suggest, polyphenol-rich diets may decrease the risk of chronic diseases by reducing oxidative stress [41]. The Fenton reaction is prevented by hydroxyl radical-scavenging flavonoids [42]. Here, the capacities of all ten plant extracts to protect against oxidative DNA damage were checked against DNA strand scission by ·OH generated in Fenton reactions on pBluescript II SK (--) DNA. We conclude that a significant contributor to DNA damage prevention is the scavenging of ·OH by the extracts of C. scariosus, C. icosandra, R. damascena and H. antidysenterica at 0.13 µg/ml, 0.16 µg/ml, 0.2 µg/ml and 0.28 µg/ml, respectively; this was corroborated by densitometric analysis.

The three effective extracts, viz. C. icosandra, R. damascena and C. scariosus, were not cytotoxic in comparison to doxorubicin, and this appears consistent with their long history of use in the Unani system of medicine.

CONCLUSIONS

Unani plants that are reported to have significant activity against several human ailments showed superior antioxidant activity as evidenced by the scavenging of the free radicals DPPH·, ABTS·+, NO, ·OH, O_2·¯ and ONOO¯. Of the ten 50% methanolic plant extracts tested, three - namely *C. icosandra, R. damascena* and *C. scariosus* - showed potentially significantly capacity to prevent oxidative DNA damage and radical scavenging activity. The *C. icosandra, R. damascena* and *C. scariosus* extracts were not cytotoxic against U937 cells. To gain further insight into the basis of their antioxidant properties, TPC, TFC and ASC were determined. All three extracts showed significantly high TPC and TFC contents, which contribute to their antioxidant activities. In conclusion, these routinely used plants can be explored further as potential sources of natural antioxidants.

AUTHORS' CONTRIBUTIONS

MDK carried out the experimental work, analyzed and interpreted the data and drafted the manuscript. DB made a substantial contribution to the conception and design of the study. AB contributed partially to the design of the study. SC supervised the work and revised the manuscript critically for important intellectual content. The authors have all read and approved the final manuscript.

ACKNOWLEDGEMENTS

This work was supported by the network research grant from the Council of Scientific and Industrial Research (CSIR). MDK acknowledges the Department of Science and Technology (DST) for her fellowship. We are grateful to acknowledge BioMedES editorial services for copyediting the manuscript. We express our gratitude to Professor Siddhartha Roy, Director, IICB, for his help and support.

REFERENCES

1. Gurib-Fakim A: Medicinal plants: Traditions of yesterday and drugs of tomorrow. *Mol Aspects Med* 2006, 27:1-93.
2. Houghton PJ: The role of plants in traditional medicine and current therapy. *J Altern Complement Med* 1995, 1:131-143.

3.   Baker D, Mocek U, Garr C: Natural products vs. combinatorials: a case study. In*Biodiversity: New Leads for the Pharmaceutical and Agrochemical Industries*. Edited by Wrigley SK, Hayes MA, Thomas R, Chrystal EJT, Nicholson N. Cambridge: The Royal Society of Chemistry; 2000:66-72.

4.   Jovanovic SV, Simic MG: Antioxidants in nutrition. *Ann NY Acad Sci* 2000, 899:326-334.

5.   Lai HY, Kim KH: *Blechnum orientale* Linn - a fern with potential as antioxidant, anticancer and antibacterial agent. *BMC Complement Altern Med* 2010, 10:15-22.

6.   Hazra B, Biswas S, Mandal N: Antioxidant and free radical scavenging activity of*Spondias pinnata*. *BMC Complement Altern Med* 2008, 8:63-72.

7.   Kumar A, Chattopadhyay S: DNA damage protecting activity and antioxidant potential of pudina extract. *Food Chem* 2007, 100:1377-1384.

8.   Ghanta S, Banerjee A, Poddar A, Chattopadhyay S: Oxidative DNA damage preventive activity and antioxidant potential of *Stevia rebaudiana* (Bertoni) Bertoni, a natural sweetener. *J Agric Food Chem* 2007, 55:10962-10967.

9.   Hertog MG, Freskens EJ, Hollman PC, Katan MB, Kromhout D: Dietary antioxidant flavonoids and risk of coronary heart disease: The Zutphen Elderly Study. *Lancet* 1993, 342:1007-1011.

10. Vickers A: Botanical medicines for the treatment of cancer: Rationale, overview of current data, and methodological considerations for Phase I and II trials. *Cancer Invest* 2002, 20:1069-1079.

11. Braca A, Sortino C, Politi M, Morelli I, Mendez J: Antioxidant activity of flavonoids from*Licania licaniaeflora*. *J Ethnopharmacol* 2002, 79:379-381.

12. Maxwell SR: Prospects for the use of antioxidant therapies. *Drugs* 1995, 49:345-361.

13. Kurup PNV: Ayurveda - A potential global medical system. In *Scientific Basis for Ayurvedic Therapies*. Edited by Mishra LC. London: CRC Press; 2004:1-15.

14. Nagulendran KR, Velavan S, Mahesh R, Begum VH: In vitro antioxidant activity and total polyphenolic content of *Cyperus rotundus* rhizomes. *E- J Chem* 2007, 4:440-449.

15. Zahin M, Aqil F, Ahmad I: The in vitro antioxidant activity and total phenolic content of four Indian medicinal plants. *Int J Pharm Pharmaceutical Sci* 2009, 1:88-95.

16. Dudonne S, Vitrac X, Coutiere P, Woillez M, Merillon JM: Comparative study of antioxidant properties and total phenolic content of 30 plant extracts of industrial interest using DPPH, ABTS, FRAP, SOD, and ORAC assays. *J Agric Food Chem* 2009, 57:1768-1774.

17. Chopra RN, Nayar SL, Chopra IC: *Glossary of Indian Medicinal Plants*. New Delhi: Council of Scientific and Industrial Research; 1956.

18. Jain SK: *Dictionary of Indian Folk Medicine and Ethnobotany*. New Delhi: Deep publications; 1991.

19. The Wealth of India: *A Dictionary of Indian Raw Materials and Industrial Products*. New Delhi: Council of Scientific and Industrial Research; 2002.

20. Darwish RM, Aburjai TA: Effect of ethnomedicinal plants used in folklore medicine in Jordan as antibiotic resistant inhibitors on *Escherichia coli. BMC Complement Altern Med* 2010, 10:9-16.

21. Yuan YV, Bone DE, Carrington MF: Antioxidant activity of dulse (*Palmaria palmata*) extract evaluated in vitro. *Food Chem* 2005, 91:485-494.

22. Zhishen J, Mengcheng T, Jianming W: The determination of flavonoid contents in mulberry and their scavenging effects on superoxide radicals. *Food Chem* 1999, 64:555-559.

23. Roe JH, Kuether CA: The determination of ascorbic acid in whole blood and urine through the 2, 4-dinitrophenylhydrazine derivatives of dihydroascorbic acid. *J Biol Chem* 1943, 147:399-407.

24. Re R, Pellegrini N, Proteggente A, Pannala A, Yang M, Rice-Evans C: Antioxidant activity applying an improved ABTS radical cation decolorization assay. *Free Radic Biol Med* 1999, 26:1231-1237.

25. Beckman JS, Chen H, Ischiropulos H, Crow JP: Oxidative chemistry of peroxynitrite. *Methods Enzymol* 1994, 233:229-240.

26. Hazra B, Sarkar R, Biswas S, and Mandal N: Comparative study of the antioxidant and reactive oxygen species scavenging properties in the extracts fruits of*Terminalia chebula, Terminalia belerica* and *Emblica officinalis. BMC Complement Altern Med* 2010, 10:20.

27. Nishikimi M, Rao AN, Yagi K: The Occurrence of Superoxide Anion in the Reaction of Reduced Phenazine Methosulfate and Molecular Oxygen. *Biochem Biophys Res Comm* 1972, 46:849-854.

28. Marcocci L, Maguire JJ, Droy-Lefaix MT, and Packer L: The nitric oxide-scavenging properties of *Ginkgo biloba* extract EGb 761. *Biochem Biophys Res Commun* 1994, 201:748-755.

29. Kim TG, Hwi KK, Hung CS: Morphological and biochemical changes of andrographolide-induced cell death in human prostatic adenocarcinoma PC-3 cells. *In vivo* 2005, 19:551-557.

30. Bravo L: Polyphenols: chemistry, dietary sources, metabolism and nutritional significance. *Nutr Rev* 1998, 56:317-333.

31. Rice-Evans CA, Miller NJ, Bramley PM, Pridham JB: The relative antioxidant activities of plant derived polyphenolic flavonoids. *Free Radic Res* 1995, 22:375-383.

32. Bors W, Heller W, Michel C, Saran M: Flavonoids as antioxidants: Determination of radical-scavenging efficiencies. *Methods Enzymol* 1990, 186:343-355.

33. Shahidi F: Natural antioxidants: An overview. In *Natural Antioxidants, Chemistry, Health Effects and Applications*. Edited by Shahidi F. Champaign: AOCS Press; 1997:1-11.

34. Beyer RE: The role of ascorbate in antioxidant protection of biomembranes: interaction with vit-E and coenzyme. *Q J Bioen Biomemb* 1994, 24:349-358.

35. Jin WY, Thuong PT, Su ND, Min BS, Son KH, Chang HW, Kim HP, Kang SS, Sok DE, Bae KH: Antioxidant Activity of Cleomiscosins A and C Isolated from *Acer okamotoanum. Arch Pharm Res* 2007, 30:275-281.

36. Ulusoy S, Bosgelmez-Tmaz G, Secilmis-Canbay H: Tocopherol, Carotene, Phenolic Contents and Antibacterial Properties of Rose Essential Oil, Hydrosol and Absolute. *Curr Microbiol* 2009, 59:554-558.

37. Schieber A, Mihalev K, Berardini N, Mollov P, Carle R: Flavonol Glycosides from Distilled Petals of *Rosa damascena* Mill. *Z Naturforsch* 2005, 60c:379-384.

38. Kumar N, Bhandari P, Singh B, Bari SS: Antioxidant activity and ultra-performance LC-electrospray ionization-quadrupole time-of-flight mass spectrometry for phenolics-based fingerprinting of Rose species: Rosa damascena, Rosa bourboniana and Rosa brunonii. *Food Chem Toxicol* 2009, 47:361-367.

39. Wei A, Shibamoto T: Antioxidant Activities and Volatile Constituents of Various Essential Oils. *J Agric Food Chem* 2007, 55:1737-1742.

40. Wiseman H, Halliwell B: Damage to DNA by reactive oxygen and nitrogen species: role in inflammatory disease and progression to cancer. *Biochem J* 1996, 313:17-29.

41. Kim HY, Kim OH, Sung MK: Effects of phenol-depleted and phenol-rich diets on blood markers of oxidative stress, and urinary excretion of quercetin and kaempferol in healthy volunteers. *J Am Coll Nutr* 2003, 22:217-223.

42. Husain SR, Cillard J, Cillard P: Hydroxyl radical scavenging activity of flavonoids. *Phytochemistry* 1987, 26:2489-2491.

Natural Antioxidant Activity of Commonly Consumed Plant Foods in India: Effect of Domestic Processing

D. Sreeramulu[1], C. V. K. Reddy[1], Anitha Chauhan[1],
N. Balakrishna[2], and M. Raghunath[1]

[1]Endocrinology, and Metabolism Division, National Institute of Nutrition, Jamai-Osmania Post Office, Tarnaka, Hyderabad, Andhra Pradesh 500007, India

[2]Statistical Division, National Institute of Nutrition, Jamai-Osmania Post Office, Tarnaka, Hyderabad, Andhra Pradesh 500007, India

ABSTRACT

Phytochemicals protect against oxidative stress which in turn helps in maintaining the balance between oxidants and antioxidants. In recent times natural antioxidants are gaining considerable interest among nutritionists, food manufacturers, and consumers because of their perceived safety, potential therapeutic value, and long shelf life. Plant foods are known to protect against degenerative diseases and ageing due to their antioxidant activity (AOA) attributed to their high polyphenolic content (PC). Data on AOA and PC of Indian plant foods is scanty. Therefore we have determined the antioxidant activity in 107 commonly consumed Indian plant foods and assessed their relation to their PC. Antioxidant activity is presented as the range of values for each of the food groups. The foods studied had good amounts of PC and AOA although they belonged to different food groups. Interestingly, significant correlation was observed between AOA (DPPH and FRAP) and PC in most of the foods, corroborating the literature that polyphenols are potent antioxidants and that they may be important contributors to the AOA of the plant foods. We have also observed that common domestic methods of processing may not affect the PC and AOA of the foods studied in general. To the best of our knowledge, these are the first results of the kind in commonly consumed Indian plant foods.

INTRODUCTION

Reactive oxygen species (ROS) such as singlet oxygen, superoxide anion, hydroxyl radical, and hydrogen peroxide (H_2O_2) are often generated as byproducts of biological reactions or from exogenous factors [1]. These reactive species exert oxidative damage by reacting with nearly every molecule found in living cells including DNA [2]. Excess ROS, if not eliminated by antioxidant system, result in high levels of free radicals and lipid peroxides which underlie the pathogenesis of degenerative diseases like atherosclerosis, carcinogenesis, diabetes, cataract, ageing, and so forth [3].

Experimental and epidemiological evidence suggests a significant role of diet in the prevention of degenerative diseases [4]. Plant derived antioxidants, such as flavonoids and related phenolic compounds, have multiple biological effects, including antioxidant activity. Phytochemicals present in plant foods exert health beneficial effects, as they combat oxidative stress in the body by maintaining a balance between oxidants and antioxidants [5]. Although more than 8000 phytochemicals have been identified in plant foods, a large percentage remains to be identified. Further, data on the polyphenolic content antioxidant activity in Indian plant foods is scanty, and the effects of domestic processing on the AOA (antioxidant activity) in Indian plant foods are not reported yet [6].

Among plant foods, green leafy vegetables and grains are a rich source of antioxidants apart from energy, protein, and selected micronutrients in Indian diets [7]. Traditionally grains and GLVs have played a major role in providing nutrition particularly in the Indian Subcontinent and in other developing countries [8]. Since plant foods are often consumed in one or the other cooked forms, polyphenol and AOA intakes calculated on the basis of their content in raw foods are likely to be inaccurate. Therefore it was considered pertinent to study the effect of domestic processing on the natural antioxidant activity and phenolic content of commonly consumed plant foods rich in these activities. Hence the effect of domestic processing (cooking) was determined on antioxidant activity and polyphenol content in some commonly consumed green leafy vegetables (GLVs) and food grains.

SAMPLING PROCEDURES

The literature on antioxidant activity and phenolic content (PC) of plant foods is limited from India as well as other parts of the world. Available literature, mostly from other parts of the world, indicates that different researchers have adopted different sampling methods to get representative value of AOA and PC. Velioglu et al. [9] collected market samples and estimated AOA and PC in 200 mgs to 1 g of fruit, vegetable, and grain products. In another study, Al-Farsi et al. [10] took 1 g of sun-dried dates to estimate antioxidant parameters whereas Arcan and Yemenicioğlu [11] took about 20 g of fresh and dry nuts for the extraction. Sampling procedures followed in some Indian studies are as follows. Gupta and Prakash [12] used one gram of green leafy Vegetable for extraction whereas Nair et al. [13] have collected fresh food samples from local market, and five grams of cleaned food sample was taken to quantify PC and flavonoids in a few Indian plant foods. Keeping in view the differences in the sampling methods used and the quantities of samples extracted by different workers to analyse antioxidants in plant foods, it appears to be a good practice to take a higher quantity of food sample for the processing specially to get reproducible results and adopt ideal sampling practices for the quantification of AOA in food and herbal samples [6].

Commonly consumed cereals, pulses, legumes, and GLVs analyzed in this study were chosen based on NNMB survey [14]. Samples were collected from market outlets located in three different locations of the twin cities of Hyderabad and Secunderabad, India. The market samples were pooled and analyzed in triplicates, and the results are presented as mean values on fresh weight basis.

To determine the effect of different types of domestic processing of grains, edible portion of the sample was sorted out and divided into four aliquots of 25 grams each [11]. First portion was processed as such to know its natural (raw) antioxidant activity, while the 2nd, 3rd, and 4th portions of the sample were subjected to conventional, pressure, and microwave methods of cooking, respectively. We have mimicked consumer's habits of food procurement from market to household. In case of GLVs 10 g edible portions were taken and processed as above.

EXTRACTION PROCEDURE

To determine the antioxidant activity in plant foods several solvent extraction procedures have been used by different researchers. There is no single satisfactory solvent extraction method suitable for the extraction of all classes of food antioxidants and phenolics. This probably is due to the differences in the chemical nature of antioxidants and phenolics, namely, simple to highly polymerized chemical substances present in plant foods. Oki et al. [15] used six different polar solvents to extract milled rice: n-hexane, diethyl ether, ethyl acetate, acetone, methanol and deionized water and found that the extracts with highly polar solvents like methanol, and deionized water shown the highest radical-scavenging activity. Al-Farsi et al. [10] used seven different solvents to extract sun-dried dates: water-phosphate buffer (40:60 ratio), methanol containing 0.1% formic acid (88:12 v/v), methanol/HCl (99.9:0.1% v/v), acetone containing 0.7% cyclodextrin, water (50:50 v/v), methanol/water (50:50 v/v), and water alone. They reported that most antioxidants present in dates were water-soluble (hydrophilic). On the other hand extraction with 50% methanol yielded the highest recovery of phenolics in the same study. This could be due to the solubility differences of phenolic acids in methanol and water. Therefore they used phosphate buffer for extracting antioxidant activity and 50% methanol to estimate total phenolic content in dates.

Rochfort and Panozzo [7] used four different solvents to extract cereal grains: (i) acetone-water (80:10 v/v), (ii) ethanol-water (80:10 v/v), (iii) methanol-water (80 : 10 v/v), and (iv) water. Found that water and 80% methanol showed higher extraction than other solvents. In another study, Chidambara Murthy et al. [16] reported that methanol extracts of grape pomace protected the activities of hepatic enzymes and could thus be important in combating reactive oxygen species. In another study using in vitro models, Singh et al. [17] observed that methanol extracts of pomegranate peel and seeds had high antioxidant activity, and similar findings were reported by others [18]. Several workers used acidified 80% methanol extraction to assess antioxidant contents in plant foods, and the reasons for it could be that methanol extraction not only gives a higher yield but also gives high antioxidant activity as compared to that with other polar solvents. Hence we have used acidic, 80% methanol (with 0.1% HCl) for extraction of phenolics and AOA from foods in our studies. Methanol extracts were also used to know the effect of domestic cooking. Domestic cooking was done with normal tap water.

Briefly, 10 grams of GLV or 25 grams of the grain sample was cooked in 100 mL of water for 10–15 minutes (in case of conventional cooking it took about 15 minutes; pressure cooking was done at 120°C for 10–12 minutes, and microwave cooking was done for 5–8 minutes, resp.). Cooking was done with the sample covered with lid except in conventional cooking. To estimate natural (raw) antioxidant content, the first portion of 10 or 25 g of the edible portion of the sample was ground in a domestic blender and extracted as such in 80% methanol containing 0.1% HCl, and final volumes of GLVs and grain samples were made to 50 and 100 mL extracts, respectively, with 80% methanol.

VARIOUS ANTIOXIDANT METHODS IN USE

Determination of AOA in plant extracts is still an unresolved problem. It is not possible to evaluate multifunctional biological antioxidants in plant foods by a single antioxidant method. Hence, batteries of tests are used; about twenty different biochemical methods are in practice to asses AOA [19]. The exact comparison of the results obtained by different methods and their general interpretation may be practically impossible due to the variability of experimental conditions and differences in physicochemical properties of oxidizable substrates. Many other factors including colloidal properties of substrate, experimental conditions, reaction medium, oxidation state, and antioxidant localization in different phases may influence antioxidant activity. Among the different antioxidant parameters in use, ABTS (Trolox equivalent antioxidant assay TEAC/ABTS) and DPPH (2, 20-Diphenyl-1-picryl hydrazyl) are widely used due to their simplicity, stability, accuracy, and reproducibility [20]. In a review on the AOA methods Huang et al. [21] suggested that FRAP (ferric reducing antioxidant power) and DPPH are the two most commonly accepted assays for the estimation of AOA in plant foods. In another study Ozgen et al. [22] evaluated the three most commonly used AOA methods and suggested that FRAP < DPPH < ABTS in fruits. A study carried out by Siddhuraju and Becker [20] suggested that DPPH < ABTS < FRAP showed better antioxidant and free radical-scavenging activities in processed cow pea and its seed extracts. Several new analytical approaches have suggested investigating antioxidant power of food extracts on the basis of their electron-donating ability. One such recently suggested assay is CAAP (chemiluminescence analysis of antioxidant power) which is a chemiluminescence based method. The rapid CAAP assay is said to be convenient to investigate the antioxidant power of herbal extracts. CAAP method showed positive correlation with FRAP (r=0.959) [23]. Nevertheless, FRAP and DPPH assays are the most widely used methods. Since these assays are electron transfer based assays and often show excellent correlation with phenolic contents, and they are carried out in acidic conditions; pH values have an important effect on the reducing capacity of antioxidants. In acidic conditions reducing capacity tends to be suppressed due to protonation on antioxidant compounds, whereas in basic conditions proton dissociation of phenolic compounds would enhance the sample reducing capacity [24].

Phenolic Content

Soluble and hydrolysable phenolic contents (free phenols) were estimated as per the procedure described by Singh et al. [17] and Singleton and Rossi [25]. Values are expressed as Gallic acid equivalents. Colorimetric method was adopted in the present study, since sensitive chromatographic method in quantification of phenols is often limited to single class of phenolics and is often limited to low-molecular weight compounds that are available as standards. It is, therefore, necessary to use colorimetric assays such as the Folin-Ciocalteu assay which rely on the reducing ability of phenols to quantify the amount of total phenolics in a sample [26].

DPPH Radical-scavenging Activity

DPPH radical-scavenging activity was determined according to Aoshima et al. [27]. This method is based on the ability of the antioxidant to scavenge the DPPH cation radical. Briefly, to 100 µL of sample extract or standard, 2.9 mL of DPPH reagent (0.1 mM in methanol) was added and vortexed vigorously. This was allowed to stand in dark for 30 min at room temperature, and the discoloration of DPPH was measured against a suitable blank at 517 nm. Percentage inhibition of the discoloration of DPPH by the sample extract was expressed as Trolox equivalents (mg/100 g).

FRAP Assay

Ferric reducing antioxidant power (FRAP) was determined according to Benzie and Strain [28]. In the presence of TPTZ, the Fe^{+2}-TPTZ complex exhibits blue color which is read at 593 nm. Briefly, 3.0 mL of working FRAP reagent was added to an appropriate volume/concentration of the sample extract, incubated for 6 min at room temperature, and the absorbance was measured at 593 nm against $FeSO_4$ standard.

OVER VIEW

Current life style is one of the major causes in the overproduction of free radicals and reactive oxygen species and decreasing physiological antioxidant capacity [29]. Food provides not only nutrients essential for life but also other bioactive compounds for health promotion and disease prevention. Epidemiological studies have consistently shown that regular consumption of plant foods is associated with reduced risk in developing chronic degenerative diseases and biological ageing [30]. Phytochemicals are the bioactive nonnutrient compounds present in plant foods which have been suggested to be responsible for their bioactivity linked to the reduced risk of major chronic diseases. It has indeed been estimated that a healthy diet could prevent approximately 30% of all cancers [31]. So far, published data from other parts of the world and India account only for a minor fraction of total polyphenols and AOA of plant foods. Therefore it was suggested to have food composition tables on antioxidant activity and polyphenolic content of commonly consumed plant foods from developing countries [32]. Hence we have attempted for the first time to get representative values of AOA in 107 commonly consumed plant foods in India. Purposive samples were purchased from three different local markets of the twin cities of Hyderabad and Secundrabad (India). They were analyzed separately and data presented on fresh weight basis, to mimic natural practice of consumption. PC and AOA were assessd in the methanol extracts of the foods by Folin-Ciocalteu method and DPPH/FRAP methods, respectively, and the results are expressed as Gallic acid and Trolox/$FeSO_4$ equivalents, respectively. It has been observed in our studies that the foods studied had good amount of polyphenols and antioxidant activity, despite the fact that they belonged to different food groups. Also, a good correlation was observed between the natural AOA of the food and it's PC (Table 2) in many of the food groups studied. Part of natural antioxidant data was published by us as full length articles; hence range of values are given. Data on the effect of domestic processing on PC and AOA has been elaborated here since it is not yet published.

NATURAL DPPH ACTIVITY IN GROUP OF PLANT FOODS

The range of DPPH activities in different food groups are presented in Table 1, and the values are expressed as mg/100 g on fresh weight basis. Among all the food groups analysed, the highest DPPH scavenging activity was observed in areca nut (28622 mg/100 g) while the activity was the least in carrots 11.06. The DPPH activity in cereals and millets ranged from 24–173 mg/100 g, with the highest activity being found in finger millet and the lowest in Semolina. The activity in legumes and pulses ranged 26–107 mg/100 g, with the highest in rajma and the lowest in roasted Bengal gram dhal. Among the nuts and oil seeds studied, the DPPH values ranged from 20–28622 mg/100, with areca nut showing the highest and coconut water having the least activities.

Among the dry fruits DPPH activity ranged 271–1541 mg/100 g, with the highest activity being in walnuts and the lowest in piyal seeds. On the other hand among fresh fruits, the values ranged 32–891 mg/100 g, with the highest in guava and the lowest in watermelon. In green leafy vegetables the values ranged 21–1021 mg/100 g, with Curry leaves having the highest whereas spinach had the least activity. In roots and tubers category the DPPH activity ranged 11–125 mg/100 g, with the highest activity being found in red beet root and the lowest in carrot. Among the vegetables studied, the DPPH values ranged 12–466 mg/100 g. The highest activity was found in okra and the lowest was in ridge gourd. Due to scanty data available in the literature on DPPH activity in Indian plant foods it was not possible for us to compare these DPPH findings with the literature.

Table 1: Natural content of AOA and TPC

S. no.	Group of foods	n(107)	Antioxidant content (mg/100 g)		PC (mg/100 g G A equ.)
			DPPH (Trol. equ.)	FRAP (FeSO$_4$ equ.)	
1	Cereals and millets	9	24–173	450–13093	47–373
2	Dry fruits	10	271–1541	1174–32416	99–959
3	Edible oils and sugars	11	3–208	11–11674	0.72–336
4	Fresh fruits	14	32–891	22–496*	26–374
5	Green leafy vegetables	11	21–1020	1380–27827	77–1077
6	Nuts and oil seeds	12	20–28622	220–4220341	10–10841
7	Pulses and legumes	11	26–107	1469–10362	62–418
8	Roots and tubers	10	11–125	256–6308	22–169
9	Vegetables	19	12–466	243–10510	27–339

Values are expressed on fresh weight basis.*ABTS: range of values are given.

Table 2: Correlation between PC versus DPPH, FRAP

S. no.	Group of Foods	n(107)	PC versus DPPH	PC versus FRAP	DPPH Versus FRAP
1	Cereals and millets	9	0.45	0.91	0.84
2	Dry fruits	10	0.97	0.87*	0.81*
3	Edible oils and sugars	14	0.93	0.93	0.99
4	Fresh fruits	14	0.77	0.84*	0.94
5	Green leafy vegetables	11	0.94	0.95	0.96
6	Nuts and oil seeds	12	0.99	0.99	0.99
7	Pulses and legumes	11	0.16	0.44	0.78
8	Roots and tubers	10	0.76	0.85	0.97
9	Vegetables	19	0.79	0.85	0.75

*ABTS: correlations are in natural form.

NATURAL FRAP ACTIVITY IN GROUP OF PLANT FOODS

The range of FRAP activities in different food groups are presented in Table 1, and the values are expressed as mg/100g on fresh weight basis. Among all the food groups analyzed, the highest FRAP activity was observed in areca nut 4220341 mg/100 g while the activity was the least in sunflower oil 36.10. Salient findings are as follows. Cereals and millets ranged 450–13093 mg/100g, highest activity was found in finger millet, and the lowest was in Semolina. Among the dry fruits, activity ranged 1174–32416 mg/100g, with the highest being in walnuts and the lowest in cashew nuts, whereas in fresh fruits, ABTS activity ranged 22–496 mg/100g, with the highest in guava and the lowest was in pineapple. Some of these findings are in agreement with the literature values of fresh fruits [32]. The AOA determined by ABTS in fresh fruits and FRAP in dry fruits showed that both had reasonably good AOA. Interestingly dry fruits had higher activity than fresh fruits probably due to their low moisture content. It was pertinent to assess whether these observations made by two different methods in fresh and dry fruits could be validated by a common, third method. Therefore, we determined the AOA in fresh and dry fruits by the DPPH scavenging method, another most commonly used antioxidant biochemical parameter. FRAP activity in green leafy vegetables ranges 1380–27827 mg/100 g, mint leaves had highest activity, and the lowest was in spinach. Edible oils and sugars range 36–11674 mg/100 g, the highest activity found in jaggery and lowest in groundnut oil (unrefined). Among the nuts and oil seeds studied, FRAP ranges 220–4220341 mg/100 g; the areca nut showed the highest activity and lowest was in coconut water. In pulses and legumes FRAP ranged 1469–10362 mg/100 g, with the highest in rajma and the lowest in green gram dhal. The roots and tubers showed a wide range 256–6308 mg/100 g, and beet root had the highest and carrot the least. Among the vegetables studied, FRAP ranges 243–10510 mg/100 g. The highest activity was found in red cabbage and the lowest in pumpkin. Due to scanty data it was not possible for us to compare our FRAP finding with the literature.

NATURAL PHENOLIC CONTENT IN GROUP OF PLANT FOODS

The soluble total phenolic content (PC) data is presented in Table 1. Values are presented as mg Gallic acid equivalent/100 g on fresh weight basis. Among all the food groups analyzed, the highest PC was observed in areca nut 10841 GAE/100 g and was the least in coconut water 10.00. Coming to different food groups, in cereals/millets PC values ranged 47–373 mg/100 g; finger millets (Ragi) had the highest (373 mg/100 g), while milled rice had the lowest (47 mg/100 g). Dry fruits values range from 99 to 959 mg/100 g of which walnuts (959 mg/100 g) and piyal seeds (99 mg/100 g) had the highest and the lowest PC, respectively. PC of fresh fruits ranged from 26 to 374 mg/100 g, with the highest PC being in guava (374) and the least in watermelon (26 mg/100 g). PC of Papaya and sapota observed here are in agreement with reported data from other parts of the world [33], orange [34], pineapple [35], and apple [36]. Among the GLVs, PC was ranging 77–1077 mg/100 g;

curry leaves have the highest (1077 mg/100 g) and the lowest was in spinach (77 mg/100 g). To compare our findings on natural phenolic contents of GLVs, as such there is very little or no published data available from India. However, Gupta and Prakash [12] analyzed phenolic content in 4 GLV samples, of which phenolic contents of fenugreek leaves values are comparable with our finding 158 versus 163 mg/100 g, whereas Amaranth and Curry leaves data is remarkably different from our findings 253 and 1077 mg/100 g, respectively, while the reported values are 150 and 387 mg/100 g. This variation could be due to the fact that they used tannic acid as standard whereas we used Gallic acid. However, it is not clear from Gupta and Prakash study [12] whether the data presented by them was on dry … or … on fresh weight basis was given. Coming to edible oils and sugars, the PC values ranged 0.72–336 mg/100 g. Jaggery had the highest PC (336 mg/100 g) while the lowest was in Vanaspati (0.72 mg/100 g). Nuts and oil seeds ranged from 10 to 10841 mg/100 g; areca nut had the highest phenolic content (10841 mg/100 g) and coconuts water the least (10 mg/100 g). Among the pulses and legumes, values ranged from 62–418 mg/100 g; black gram dhal had the highest (418 mg/100 g) while green gram dhal had the least (62 mg/100 g). Roots and tubers showed a wide range (22–169 mg/100 g), and beet root had the highest and carrot the least. Phenolic content of vegetables ranged from 27 to 339 mg/100 g, and red cabbage had the highest and ridge gourd the lowest. Very little published data is available on PC of Indian plant foods; some findings are in agreement with our data [37]. However, phenolic contents of plant foods can significantly vary due to various other factors, like plant genetics and cultivar, soil composition and growing conditions, maturity state and postharvest conditions, and so forth [38].

CORRELATION BETWEEN PC, DPPH, AND FRAP (IN A GROUP OF NATURAL PLANT FOODS)

Our observations on correlation between DPPH, FRAP, and PC of cereals, pulses, and legumes are in agreement with an earlier report [18] in that no significant correlation was observed between these two parameters among these food grains. Interestingly, no correlation was observed among PC, DPPH, and ABTS in wheat extracts [20]. The lack of correlation in cereal and legume grains could be due to the differences in the bound and free forms of phytochemicals present in them. It was observed that there was a possibility of underestimation of phenolic compounds in cereal/legume grains due to the differences in bound and free phenolics present in them. The bound phenolics contribute about 62% in rice to 85% in corn [5]. Another possibility could be due to the different responses of different phenolic compounds in different assay systems. Since the molecular antioxidant responses of phenolic compounds vary remarkably, depending on their chemical structure, their AOA does not necessarily correlate with the PC in grains [39].

However, both in dry and fresh fruits, there was a good correlation between PC and AOA, and our findings are in agreement with the available literature on the phenolic content of fresh and dry fruits [10]. However the discordance in phenolic content of different groups of foods studied could be due to varietal, seasonal, agronomical, and genomic differences, moisture content, method of extraction and standards used, and so forth [40]. Among GLVs, there was a good correlation among the PC and antioxidant parameters studied (Table 2). However, little information is available in the literature on the AOA and PC correlations in GLVs [12].

Although edible oils and sugars, belong to different food groups, there was a good correlation among their PC and AOA parameters in that the "r" value was 0.93 between both the AOA parameters and PC. Among nuts and oil seeds, a significant correlation was observed between AOA (both DPPH and FRAP) and PC. The "" value between PC and AOA was 0.99, indicating the importance of PC to their AOA as assessed by these two methods. These findings are in agreement with earlier reports of this nature [10].

Correlations between the antioxidant activity and phenolic content of roots, tubers, and vegetables are given in Table 2. In general, there was a good correlation between the PC and AOA among the vegetables, roots, and tubers studied. A significant correlation ($p<0.01$) was observed between PC and AOA both in roots and tubers (r values being 0.76 and 0.85, resp. with DPPH and FRAP) and other vegetables (r=0.79and 0.85 with DPPH and FRAP). These findings suggest that PC may be an important contributor to the AOA of roots, tubers, and vegetables; our observations are in agreement with the literature from other parts of the world [34].

EFFECT OF DOMESTIC COOKING ON PC AND AOA IN GREEN LEAFY VEGETABLES

Plant foods are often consumed in one or the other processed forms. Therefore, it was considered pertinent to study the effect of common domestic processing (cooking) methods on the natural antioxidant activity and phenolic content of a few commonly consumed plant foods. Since oxidants and antioxidants have different chemical and physical characteristics, different types of cooking may bring different type of alterations in antioxidant activities of different foods. Further, if polyphenol intakes are calculated based on raw plant foods, the intake values computed may not be accurate. Hence effect of cooking was determined on phenolic content and antioxidant activity in commonly consumed green leafy vegetables (GLVs) and food grains.

Effect of cooking on PC and AOA of GLVs is presented in Tables 3–6. Phenolic content and antioxidant activity of foods cooked by different methods were compared with its natural (raw) activity from same portion of subsample. In general different cooking methods used in this study did not affect the AOA and phenolic contents in most of the GLVs. Percent change in the phenolic content (PC) or antioxidant activity (AOA) on cooking is given in parentheses with respective raw GLV value (Tables 3–6). Differences were considered significant at a value at least <0.05.

Table 3: Effect of domestic processing on polyphenol content of commonly consumed green leafy vegetables

Sl. no.	Common name	Botanical name	Phenolic content (mg/100 g Gallic acid Eq.)			
			Raw	Conventional	Pressure	Micro-wave
1	Amaranth	Amaranthus gangeticus	253.0[a] (100)	275[b] (108)	355[c] (140)	312[d] (123)
2	Ambat chukka	Rumex vesicarius	100.3 (100)	90 (89)	93 (92)	91 (91)

3	Coriander leaves	Coriandrum sativum	239.6ᵃ (100)	417ᵇ (174)	451ᶜ (188)	506ᵈ (211)
4	Curry leaves	Murraya koenigii	1077.0ᵃ (100)	1434ᵇ (133)	1184ᶜ (109)	1377ᵈ (127)
5	Fennel leaves	Foeniculum vulgare	251.3 (100)	268 (106)	265 (105)	312 (124)
6	Fenugreek leaves	Trigonella foenumgraecum	163.3ᵃ (100)	180ᵃ (110)	176ᵃ (107)	220ᵇ (134)
7	Purslane leaves	Portulaca oleracea	94.6ᵃ (100)	128ᵇ (135)	138ᵇ (146)	128ᵇ (135)
8	Sorrel leaves	Hibiscus cannabinus	191.3 (100)	194 (101)	211 (107)	213 (111)
9	Mint	Mentha spicata	440.3ᵃ (100)	657ᵇ (149)	796ᶜ (180)	761ᶜ (172)
10	Water amaranth	Alternanthera sessilis	136.3 (100)	122 (89)	110 (80)	123 (90)
11	Spinach	Spinacia oleracea	77.0ᵃ (100)	96ᵇ (125)	125ᶜ (162)	117ᶜ (152)

Mean values were compared (n=3) by nonparametric Kruskal Wallis one way ANOVA. Differences in alphabets are significantly different at $P < 0.05$. Percent gain or loss calculated when raw value taken as 100%. Percent recovery values are given in parentheses. Decimal points are not given due to higher numbers.

Table 4: Effect of domestic processing on DPPH activity of commonly consumed green leafy vegetables

Sl. no.	Common name	Botanical name	DPPH (mg/100 g Trolox Eq.)			
			Raw	Conventional	Pressure	Microwave
1	Amaranth	Amaranthus gangeticus	405.6ᵃ (100)	520ᵇ (128)	527ᵇ (129)	476ᵇ (117)
2	Ambat chukka	Rumex vesicarius	85.3 (100)	87 (101)	83 (97)	94 (110)
3	Coriander leaves	Coriandrum sativum	471.0ᵃ (100)	886ᵇ (181)	948ᵇ (201)	1100ᶜ (233)
4	Curry leaves	Murraya koenigii	1020.6ᵃ (100)	950ᵇ (93)	1724ᶜ (168)	1418ᵈ (138)

5	Fennel leaves	Foeniculum vulgare	545.3 (100)	592 (108)	540 (99)	746 (136)
6	Fenugreek leaves	Trigonella foenum-graecum	144.3 (100)	142 (98)	127 (87)	193 (134)
7	Purslane leaves	Portulaca oleracea	138.3 (100)	162 (117)	165 (119)	151 (109)
8	Gogu	Hibiscus cannabinus	346.0 (100)	365 (105)	334 (96)	456 (131)
9	Mint	Mentha spicata	1368.6 (100)	2055 (150)	1856 (135)	2020 (147)
10	Ponna-ganti	Alternan-thera sessilis	173.0 (100)	172 (99)	203 (117)	198 (114)
11	Spinach	Spinacia oleracea	21.6^a (100)	69^b (321)	85^c (393)	104^d (481)

Mean values were compared (n=3) by nonparametric Kruskal wallis one way ANOVA. Differences in alphabets are significantly different at $P < 0.05$. Percent gain or loss calculated when raw value taken as 100%. Percent recovery values are given in parentheses. Decimal points are not given due to higher numbers.

Table 5: Effect of domestic processing on FRAP activity of commonly consumed green leafy vegetables

Sl. no.	Common name	Botanical name	FRAP (mg/100 g FeSO$_4$ Eq.)			
			Raw	Conventional	Pressure	**Microwave**
1	Amaranth	Amaranthus gangeticus	8237.6^a (100)	11370^b (138)	12102^b (146)	11786^b (143)
2	Ambat chukka	Rumex vesicarius	3511.6 (100)	3270 (93)	2946 (83)	3243 (92)
3	Coriander leaves	Coriandrum sativum	7125.6^a (100)	18636^b (261)	16123^c (226)	19802^d (277)
4	Curry leaves	Murraya koenigii	20275.0^a (100)	18533^b (91)	24213^c (119)	27392^d (135)
5	Fennel leaves	Foeniculum vulgare	9238.6^a (100)	10128^a (109)	9970^a (107)	13362^b (144)
6	Fenugreek leaves	Trigonella foenum-graecum	3409.6^a (100)	3919^b (114)	4799^c (140)	5429^d (159)
7	Purslane leaves	Portulaca oleracea	2863.3^a (100)	4327^b (151)	4800^c (167)	4030^b (140)
8	Gogu	Hibiscus cannabinus	5254.0 (100)	7274 (138)	6921 (131)	7107 (135)
9	Mint	Mentha spicata	27827.6^a (100)	42562^b (152)	$48909^{b,c}$ (175)	50401^c (181)

| 10 | Ponnaganti | Alternanthera sessilis | 5068.3 (100) | 4280 (84) | 4837 (95) | 4327 (85) |
| 11 | Spinach | Spinacia oleracea | 1380.6[a] (100) | 3196[b] (231) | 3471[b] (251) | 3502[b] (253) |

Mean values were compared (n=3) by nonparametric Kruskal wallis one way ANOVA. Differences in alphabets are significantly different at $P < 0.05$. Percent gain or loss calculated when raw value taken as 100%. Percent recovery values are given in parentheses. Decimal points are not given due to higher numbers.

Table 6: Rank correlation between phenolic content versus DPPH and FRAP in different cooking methods of GLV

TPC versus AOA	Raw	**Traditional**	**Pressure**	**Microwave**	**Homogeneity**
TPC versus DPPH	0.945	0.936	0.918	0.945	$\chi2 = 0.23$, $P=0.97$
TPC versus FRAP	0.955	0.936	0.927	0.973	$\chi2 = 1.23$, $P=0.74$
DPPH versus FRAP	0.964	0.973	0.991	0.991	$\chi2 = 3.23$, $P=0.36$

All correlations are significant at $p<0.001$ (n=11).

Out of the eleven GLVs studied, only two GLVs, namely, Ambat Chukka and Ponnaganti showed a small decrease of 10–20% in their PC content on cooking. While Gogu showed very little or no change on cooking. The other eight GLVs showed an increase in PC during different types of cooking (Table 3). Among them six GLVs, namely, Amaranth, Curry leaves, Fennel, Fenugreek, Purslane and Sorrel leaves showed an increase in PC, ranging 108–146% on cooking. Coriander, Mint, and spinach showed a significant increase in PC in the above cooking methods, and the increase was ranging 125–211% (Table 1). As such there is very little data of this kind reported in the literature. Kuti and Konuru [41] demonstrated in spinach leaves a similar increasing trend in PC on cooking. Contrary to this, Faller and Fialho [42] showed cooking loss in PC in vegetables. The possible explanation for the increasing or decreasing trends of phenolic contents during various cooking methods could be that the phenolics are stored in pectin or cellulose networks of plant foods and can be released during thermal processing. In turn individual phenolics May sometime increase because heat can break supramolecular structure which might make the phenolic compounds react better with the reagents [43].

Effect of cooking on DPPH scavenging activity is given in Table 4. The increase or decrease activity in different GLVs during different cooking methods was compared with its natural DPPH activity of the raw GLV. Our findings on changes in DPPH are in line with those in PC. In general, an increasing trend in DPPH activity on cooking was seen in most of the GLVs studied. Out of eleven GLVs, marginal effect of 1–10% was seen only in Ambat Chukka. Most of the GLVs showed an increasing trend in all the three methods of cooking. Among them Purslane and Ponnaganti showed 10–20% increase, whereas Amaranth and Mint showed 17–50% increase. Coriander and Curry leaves showed an increase of 38–133%. During conventional cooking, curry leaves showed little effect (<7%) while spinach showed an enormous increase of 221–381%. Remaining three GLVs namely, Fennel, Fenugreek, and Gogu leaves did not show any effect in conventional and pressure cooking but in microwave cooking alone showed about 31–36% increase (Table 4). This could be due to effect of high temperature as compared to the above two methods of heat treatment. Considering that no data of similar type is available from other parts of the world, we are unable to compare our findings with literature reports. Data available on other vegetables (not GLVs) reported a mixed trend, which is in agreement with our results. Indeed an increasing trend was observed in potatoes [44], while a decreasing trend was reported in other vegetables [42].

Effect of cooking on FRAP is presented in Table 5. This biochemical indicator was chosen as the second most commonly used antioxidant biochemical parameter. Again, similar increasing trends were seen in FRAP activity on cooking. Most of the GLVs (nine out of eleven) showed an increase, ranging 119–181%. While Coriander and spinach leaves showed an enormous increase, two other GLVs, Ambat Chukka and Ponnaganti, showed a decrease (maximum of 10%). This type of complex trend during cooking requires further research [45]. However, the present data on natural antioxidant content in commonly consumed GLVs is the first data of its kind from India. Secondly our findings on the effect of different methods of cooking in above GLVs, most of them, show an increase in AOA; it could be due to better availability of bound phenolics. Correlation among the three biochemical parameters and effect of different methods of cooking were assessed next, and these correlations were compared by using rank correlations (Table 6). Phenolics versus antioxidant parameters in different cooking methods are highly correlated. Findings of this study suggest that although different cooking methods showed changes (highly significant in some cases) in the phenolic content and AOA of the GLVs, there was no effect of domestic cooking on the correlation between the PC and AOA. This observation confirms that PC may be important contributor to the AOA of GLVs both in raw and cooked forms.

EFFECT OF DOMESTIC COOKING (FOOD GRAINS)

PC and AOA of the food grains (raw and cooked by different methods) are presented in Tables 7–9. The PC of raw whole green gram (with peel) was the highest (284 mg/100 g) followed by black rajma (146 mg/100 g). Green gram dhal without peel had the least phenolic content (41 mg/100 g) (Table 7). This difference in phenolic content of green gram whole and dhal could be due to the peel component, known to contribute high phenolic contents in grains. DPPH scavenging activity was the highest in black rajma (160 mg/100 g) followed by whole green gram (113 mg/100 g), and the lowest was in green gram dhal (without peel) (21 mg/100 g) (Table 8). FRAP content was the highest in black rajma followed by soya bean and the lowest was in green gram dhal. The FRAP values were 6852, 3778, and 1066 mg/100 g, respectively (Table 9).

Table 7: Effect of domestic processing on polyphenol content of commonly consumed pulses and legumes in India

Sl. No.	Common name	Botanical name	Phenolic content (mg/100 g Gallic acid Eq.)				P value
			Raw	Conventional	Pressure	Microwave	
1	Bengal gram dhal	Cicer arietinum	92.6 ± 5.5[a] (100)	90.6 ± 6.5[a] (98)	98.6 ± 4.0[a] (106)	86.0 ± 5.5[a] (93)	NS
2	Bengal gram dhal (roasted)	Cicer arietinum	116.3 ± 7.7[a] (100)	105.6 ± 6.1[a] (91)	108.6 ± 5.6[a] (93)	102.0 ± 10.5[a] (88)	NS

3	Bengal gram (whole grains)	Cicer arietinum	114.0 ± 10.4^a (100)	154.6 ± 7.0^b (136)	176.3 ± 4.5^c (154)	113.3 ± 6.0^d (99)	0.024
4	Black gram dhal (without peel)	Phaseolus mungo Roxb	69.3 ± 4.5^a (100)	58.6 ± 3.0^b (85)	60.0 ± 2.6^b (86)	51.3 ± 3.2^c (74)	0.022
5	Green gram dhal	Phaseolus aureus Roxb	41.3 ± 2.5^a (100)	43.6 ± 1.1^a (106)	43.0 ± 3.6^a (104)	34.0 ± 3.0^c (82)	NS
6	Green gram dhal (whole)	Phaseollus aureus Roxb	284.3 ± 6.5^a (100)	249.3 ± 3.0^b (88)	269.3 ± 4.5^c (95)	243.6 ± 4.0^b (86)	0.019
7	Lentil	Lens esculenta	64.3 ± 2.5^a (100)	64.6 ± 3.5^a (100)	59.0 ± 6.0^a (92)	56.0 ± 2.6^a (87)	NS
8	Peas green (dry)	Pisum sativum	82.3 ± 2.0^a (100)	84.0 ± 2.6^a (102)	103.3 ± 5.5^b (126)	75.6 ± 3.5^c (92)	0.024
9	Red gram dhal (without peel)	Cajanus cajan	70.0 ± 6.5^a (100)	83.6 ± 4.6^b (119)	81.6 ± 1.5^b (117)	74.0 ± 4.5^a (106)	0.035
10	Rajma (Black)	Phaseolus Vulgaris	146.6 ± 7.0^a (100)	186.0 ± 4.5^b (127)	195.6 ± 9.7^c (133)	159.3 ± 2.5^c (109)	0.020
11	Soya bean	Glycine max Merr.	81.6 ± 3.5^a (100)	82.0 ± 7.5^a (100)	98.3 ± 5.0^a (121)	94.3 ± 6.0^a (116)	NS

Pooled samples were analyzed in triplicates. Data is presented as mean ± SD. Mean values were compared by nonparametric Kruskal Wallies H test of one way ANOVA. Differences in alphabets are significantly different at P < 0.05. Percent gain or loss calculated when raw value taken as 100%. Percent recovery values are given in parenthesis. Decimal points are not given due to higher numbers.

Table 8: Effect of domestic processing on DPPH activity of commonly consumed Pulses and Legumes in India

Sl. no.	Common name	Botanical name	DPPH (mg/100g Trolox Eq.)				P value
			Raw	Conventional	Pressure	Microwave	
1	Bengal gram dhal	Cicer arietinum	42.6 ± 2.5[a] (100)	43.3 ± 1.5[a] (102)	43.6 ± 4.0[a] (102)	40.0 ± 3.6[a] (94	NS
2	Bengal gram dhal (roasted)	Cicer arietinum	31.3 ± 2.5[a] (100)	34.3 ± 3.7[a] (110)	31.3 ± 3.5[a] (100)	25.6 ± 2.5[a] (82)	NS
3	Bengal gram (whole grains)	Cicer arietinum	68.6 ± 4.5[a] (100)	100.0 ± 7.5[b] (146)	95.3 ± 3.5[b] (139)	60.3 ± 2.5[c] (88)	0.022
4	Black gram dhal (with out peel)	Phaseolus mungo Roxb	35.0 ± 3.0[a] (100)	29.0 ± 1.0[a] (83)	30.0 ± 7.2[a] (86)	24.6 ± .0[a] (70)	NS
5	Green gram dhal	Phaselus aureus Roxb	21.3 ± 4.5[a] (100)	19.3 ± 4.5[a] (91)	17.6 ± 3.5[a] (83)	18.6 ± 3.6[a] (87)	NS
6	Green gram dhal (whole)	Phaseolus aureus Roxb	113.6 ± 9.2[a] (100)	184.3 ± 9.0[b] (162)	159.3 ± 13.7[c] (140)	171.3 ± 9.0[b.c] (151)	0.027
7	Lentil	Lens esculenta	35.6 ± 3.5[a] (100)	38.0 ± 4.0[a] (107)	35.3 ± 3.7[a] (99)	36.6 ± 6.5[a] (103)	NS
8	Peas green (dry)	Pisum sativum	51.0 ± 3.0[a] (100)	55.3 ± 3.0[a] (108)	56.0 ± 4.0[a] (110)	42.0 ± 5.5[b] (82)	0.040
9	Red gram dhal (without peel)	Cajanus cajan	42.0 ± 4.0[a] (100)	49.3 ± 7.5[a] (117)	56.3 ± 4.7[a] (134)	42.0 ± 4.0[a] (100)	NS
10	Rajma (Black)	Phaseolus Vulgaris	160.0 ± 8.1[a] (100)	182.3 ± 4.5[a] (114)	170.3 ± 6.0[a] (106)	174.0 ± 9.5[a] (109)	NS
11	Soya been	Glycine max Merr.	75.6 ± 7.5[a] (100)	61.3 ± 2.3[b] (81)	59.3 ± 4.1[c] (78)	71.6 ± 2.0[a] (95)	0.023

Pooled samples were analysed in triplicates. Data is presented as mean ± SD. Mean values were compared by non-parametric Kruskal Wallies H test of one way ANOVA. Differences in alphabets are significantly different at P < 0.05. Percent gain or loss calculated when raw value taken as 100%. Percent recovery values are given in parenthesis. Decimal points are not given due to higher numbers.

Table 9: Effect of domestic processing on FRAP activity of commonly consumed Pulses and Legumes in India

S. no.	Common name	Botanical name	FRAP (mg/100 g $FeSO_4$ Eq)				P Value
			Raw	Conventional	Pressure	Microwave	
1	Bengal gram dhal	Cicer arietinum	1679 ± 53.2[a] (100)	1909 ± 64.7[a] (114)	1968 ± 44.1[a] (117)	1973 ± 46.6[a] (118)	NS
2	Bengal gram dhal (roasted)	Cicer arietinum	1466 ± 125.2[a] (100)	1711 ± 109.5[a] (117)	1359 ± 114.5[a] (93)	1367 ± 103.5[a] (93)	NS
3	Bengal gram (whole grains)	Cicer arietinum	2283 ± 132.8[a] (100)	2560 ± 131.0[b] (112)	2676 ± 170.0[b] (117)	2177 ± 102.1[a] (95)	0.033
4	Black gram dhal (without peel)	Phaseolus mungo Roxb	1515 ± 41.4[a] (100)	1420 ± 80.1[a] (94)	1470 ± 46.5[a] (97)	1265 ± 47.8[a] (83)	NS
5	Green gram dhal	Phaseolus aureus Roxb	1066 ± 128.6[a] (100)	1371 ± 58.3[a] (128)	1042 ± 99.8[a] (98)	938 ± 85.7[a] (88)	NS
6	Green gram dhal (whole)	Phaseolus aureus Roxb	3098 ± 22.4[a] (100)	5490 ± 101.0[b] (177)	5785 ± 184.6[c] (187)	5505 ± 81.1[b] (178)	0.025
7	Lentil	Lens esculenta	1534 ± 54.0[a] (100)	1652 ± 121.0[a] (108)	2058 ± 109.0[a] (134)	1625 ± 107.9[a] (105)	NS
8	Peas green (dry)	Pisum sativum	1846 ± 80.8[a] (100)	3027 ± 93.7[a] (164)	3734 ± 71.0[b] (202)	2609 ± 64.5[c] (141)	0.016
9	Red gram dhal (without peel)	Cajanus cajan	2446 ± 84.9[a] (100)	3133 ± 81.6[b] (128)	4251 ± 106.6[c] (173)	2646 ± 84.8[b] (108)	0.016

10	Rajma (Black)	Phaseolus Vulgaris	6852 ± 66.4[a] (100)	6809 ± 125.2[a] (99)	7171 ± 81.4[b] (105)	7915 ± 130.5[c] (115)	0.025
11	Soya been	Glycine max Merr.	3778 ± 162.5[a] (100)	3504 ± 128.0[a] (93)	3714 ± 125.5[a] (98)	3502 ± 149.0[a] (93)	NS

Pooled samples were analysed in triplicates. Data is presented as mean ± SD. Mean values were compared by nonparametric Kruskal Wallies test of one way ANOVA. Differences in alphabets are significantly different at $P < 0.05$. Percent gain or loss calculated when raw value taken as 100%. Percent recovery values are given in parenthesis. Decimal points are not given due to higher numbers.

Effect of different cooking methods on antioxidant activity of each food grain was compared with its AOA and phenolic contents of the raw sample. (Tables 7–9). Percentage change in the PC and antioxidant activity on cooking is given in parentheses in Tables 7–9. Overall, different cooking methods did not show any significant cooking losses but showed mixed results of increasing and/or decreasing trends (Tables 7–9), the changes being significant in most of the whole grains as compared to grains without seed coat.

Effects of cooking on PC are presented in Table 7. Nine out of 11 legumes samples showed the maximum of 20% increase or decrease in their PC during different types of domestic cooking. Interestingly, during conventional and pressure cooking, whole Bengal gram and rajma have shown 27 and 54% increase. Other studies showed similar effects on AOA in potatoes upon cooking [44] and in other vegetables [38]. The possible mechanism for the increase or decrease in AOA during various cooking methods could be that the phenolics were stored in pectin or cellulose networks of plant foods and were released during thermal processing [39].

DPPH scavenging activity in legumes cooked by different cooking methods also showed a mixed/inconsistent trend (Table 8). Nine out of eleven food grains studied showed less than 20% increase or decrease during cooking. It is however interesting that whole green gram (with peal) showed a higher increase in DPPH activity in all cooking methods studied, with the increase ranging 40–62% as compared to its content in the unprocessed form. Indeed some literature says that this type of complex trend on cooking is unexplainable and requires further research [45].

Effect of cooking on FRAP activity is given in Table 9. Findings are in line with DPPH, showing a mixed trend. Nine out of eleven legumes showed less than 20% variation in FRAP values. While whole green gram and dry green peas showed higher increase in FRAP ranging 41–102% in different methods of cooking; lentil and red gram dhal showed 34–73% increase albeit during pressure cooking only. It was however of interest that over all the percent increase or decrease found vis-à-vis their content in unprocessed food showed similar trend in different cooking methods in a given food grain. Such increasing or decreasing trends were reported in few vegetables from other parts of the world [46]. The possible explanation given for this type of finding was summarized by few workers as follows. Cooking could have resulted in liberation of high amounts of antioxidant compounds due to thermal destruction of cell wall and subcellular compartments [47, 48]. Another possibility might be the production of stronger radical-scavenging antioxidants by thermal or chemical reactions [49]. There can be a production of new nonnutritional antioxidants or formation of novel compounds such as Millard reaction products with antioxidant activity during cooking. However these findings are first of their kind in commonly consumed pulses and legumes.

Correlations among the PC and AOA were determined in the legumes in unprocessed as well as during the three different types of domestic cooking. For this purpose rank correlations were used, and the data is presented in Table 10. Correlations between PC and AOA were significant in different cooking methods, and they were comparable across the methods. Although different cooking methods showed changes (highly significant in some cases) in the phenolic content and AOA of the food grains, the finding that they did not affect the correlation between the PC and AOA suggests that PC may be important contributor to the AOA even in pulses and legumes, both in raw and cooked forms.

Table 10: Rank correlation between phenolic content and AOA (DPPH and FRAP) in raw and cooked pulses and legumes

TPC versus AOA	Raw	Traditional	Pressure	Microwave	Homogeneity
TPC versus DPPH	0.689	0.801	0.793	0.780	$\chi2$= 1.23, P = 0.746
TPC versus FRAP	0.573	0.701	0.619	0.706	$\chi2$= 1.12, P = 0.772
DPPH versus FRAP	0.918	0.909	0.895	0.916	$\chi2$= 0.31, P = 0.959

All correlations are significant at $P < 0.01$ (n=11), and correlations are comparable across the methods. Between the methods, all the parameters are significantly correlated (TPC versus DPPH, TPC versus FRAP, and DPPH versus FRAP).

CONCLUSIONS

To the best of our knowledge, findings observed in this review are first of their kind from India, this review mainly dealt with two aspects, and natural antioxidant content of commonly consumed plant foods in India was assessed and correlated with its phenolic content. And the second aspect is assessing the effect of domestic cooking on PC and antioxidant activity for the first time from India in the most commonly consumed GLVs and grains. Our findings demonstrate that antioxidant contents did not get affected in most of the foods studied; on the other hand most of them shown a higher AOA in different method of domestical processing. This overview would be useful to researchers, nutritionists, and consumers to assess AOA and/ or formulate antioxidant-rich therapeutic diets as well as commercial antioxidant-rich preparations from plant foods. In addition, they will be a valuable addition to the scanty knowledge on antioxidant activity of commonly consumed foods in India.

Limitation of the Present Findings: Purposive samples were collected from three local markets to provide first-hand information on antioxidant activity of plant foods commonly consumed in India. Hence, findings cannot be considered as Indian plant foods data base.

ACKNOWLEDGMENTS

The authors thank Dr. Kalpagam Polasa, Officer in charge, National Institute of Nutrition, for her encouragement. They thank Mr. R. Srinivas Rao for his help in preparation of the paper.

REFERENCES

1.   H. Wiseman and B. Halliwell, "Damage to DNA by reactive oxygen and nitrogen species: role in inflammatory disease and progression to cancer," Biochemical Journal, vol. 313, no. 1, pp. 17–29, 1996.

2.   M. S. Cooke, M. D. Evans, M. Dizdaroglu, and J. Lunec, "Oxidative DNA damage: mechanisms, mutation, and disease," The FASEB Journal, vol. 17, no. 10, pp. 1195–1214, 2003

3.   B. Halliwell, J. M. C. Gutteridge, and C. E. Cross, "Free radicals, antioxidants, and human disease: where are we now?" Journal of Laboratory and Clinical Medicine, vol. 119, no. 6, pp. 598–620, 1992.

4.   P. J. Harris and L. R. Ferguson, "Dietary fibre: its composition and role in protection against colorectal cancer," Mutation Research, vol. 290, no. 1, pp. 97–110, 1993.

5.   A. Scalbert, C. Manach, C. Morand, C. Rémésy, and L. Jiménez, "Dietary polyphenols and the prevention of diseases," Critical Reviews in Food Science and Nutrition, vol. 45, no. 4, pp. 287–306, 2005.

6.   H. Zieliński and H. Kozłowska, "Antioxidant activity and total phenolics in selected cereal grains and their different morphological fractions," Journal of Agricultural and Food Chemistry, vol. 48, no. 6, pp. 2008–2016, 2000

7.   S. Rochfort and J. Panozzo, "Phytochemicals for health, the role of pulses," Journal of Agricultural and Food Chemistry, vol. 55, no. 20, pp. 7981–7994, 2007.

8.   D. Sreeramulu, C. V. K. Reddy, and M. Raghunath, "Antioxidant activity of commonly consumed cereals, millets, pulses and legumes in India," Indian Journal of Biochemistry and Biophysics, vol. 46, no. 1, pp. 112–115, 2009

9. Y. S. Velioglu, G. Mazza, L. Gao, and B. D. Oomah, "Antioxidant activity and total phenolics in selected fruits, vegetables, and grain products," Journal of Agricultural and Food Chemistry, vol. 46, no. 10, pp. 4113–4117, 1998.

10. M. Al-Farsi, C. Alasalvar, A. Morris, M. Baron, and F. Shahidi, "Comparison of antioxidant activity, anthocyanins, carotenoids, and phenolics of three native fresh and sun-dried date (Phoenix dactylifera L.) varieties grown in Oman," Journal of Agricultural and Food Chemistry, vol. 53, no. 19, pp. 7592–7599, 2005.

11. I. Arcan and A. Yemenicio□lu, "Antioxidant activity and phenolic content of fresh and dry nuts with or without the seed coat," Journal of Food Composition and Analysis, vol. 22, no. 3, pp. 184–188, 2009.

12. S. Gupta and J. Prakash, "Studies on Indian green leafy vegetables for their antioxidant activity," Plant Foods for Human Nutrition, vol. 64, no. 1, pp. 39–45, 2009.

13. S. Nair, R. Nagar, and R. Gupta, "Antioxidant phenolics and flavonoids in common Indian foods,"Journal of Association of Physicians of India, vol. 46, no. 8, pp. 708–710, 1998.

14. National Nutrition Monitoring Bureau, Report of Repeat Surveys (1988–90), National Institute of Nutrition, ICMR, Hyderabad, India, 1991.

15. T. Oki, M. Masuda, M. Kobayashi et al., "Polymeric procyanidins as radical-scavenging components in red-hulled rice," Journal of Agricultural and Food Chemistry, vol. 50, no. 26, pp. 7524–7529, 2002.

16. K. N. Chidambara Murthy, R. P. Singh, and G. K. Jayaprakasha, "Antioxidant activities of grape (Vitis vinifera) pomace extracts," Journal of Agricultural and Food Chemistry, vol. 50, no. 21, pp. 5909–5914, 2002.

17. R. P. Singh, K. N. Chidambara Murthy, and G. K. Jayaprakasha, "Studies on the antioxidant activity of pomegranate (Punica granatum) peel and seed extracts using in vitro models," Journal of Agricultural and Food Chemistry, vol. 50, no. 1, pp. 81–86, 2002.

18. B. Matthäus, "Antioxidant activity of extracts obtained from residues of different oilseeds," Journal of Agricultural and Food Chemistry, vol. 50, no. 12, pp. 3444–3452, 2002.

19.   P. Stratil, B. Klejdus, and V. Kubáň, "Determination of total content of phenolic compounds and their antioxidant activity in vegetables—evaluation of spectrophotometric methods," Journal of Agricultural and Food Chemistry, vol. 54, no. 3, pp. 607–616, 2006.

20.   P. Siddhuraju and K. Becker, "The antioxidant and free radical scavenging activities of processed cowpea (Vigna unguiculata (L.) Walp.) seed extracts," Food Chemistry, vol. 101, no. 1, pp. 10–19, 2007.

21.   D. Huang, B. Ou, and R. L. Prior, "Review on AOA methods: the chemistry behind antioxidant capacity assays," Journal of Agricultural and Food Chemistry, vol. 53, no. 6, pp. 1841–1856, 2005.

22.   M. Ozgen, R. N. Reese, A. Z. Tulio Jr., J. C. Scheerens, and A. R. Miller, "Modified 2,2-azino-bis-3-ethylbenzothiazoline-6-sulfonic acid (ABTS) method to measure antioxidant capacity of selected small fruits and comparison to ferric reducing antioxidant power (FRAP) and 2,2'-diphenyl-1- picrylhydrazyl (DPPH) methods," Journal of Agricultural and Food Chemistry, vol. 54, no. 4, pp. 1151–1157, 2006.

23.   T. Chen, S. Liou, H. Wu et al., "New analytical method for investigating the antioxidant power of food extracts on the basis of their electron-donating ability: comparison to the ferric reducing/antioxidant power (FRAP) assay," Journal of Agricultural and Food Chemistry, vol. 58, no. 15, pp. 8477–8480, 2010.

24.   R. Re, N. Pellegrini, A. Proteggente, A. Pannala, M. Yang, and C. Rice-Evans, "Antioxidant activity applying an improved ABTS radical cation decolorization assay," Free Radical Biology and Medicine, vol. 26, no. 9-10, pp. 1231–1237, 1999.

25.   V. L. Singleton and J. A. Rossi, "Colorimetry of total phenolics with phosphomolybdic phosphotungstic acid reagents," American Journal of Enology and Viticulture, vol. 16, pp. 144–158, 1965.

26.   I. Johnson and G. Williamson, Eds., Phytochemical Functional Foods, Woodhead Publication in Food Science and Technology, CRC Publication, 2007.

27.   H. Aoshima, H. Tsunoue, H. Koda, and Y. Kiso, "Aging of whiskey increases 1,1-diphenyl-2-picrylhydrazyl radical scavenging

activity," Journal of Agricultural and Food Chemistry, vol. 52, no. 16, pp. 5240–5244, 2004.

28. I. F. F. Benzie and J. J. Strain, "Ferric reducing/antioxidant power assay: direct measure of total antioxidant activity of biological fluids and modified version for simultaneous measurement of total antioxidant power and ascorbic acid concentration," Methods in Enzymology, vol. 299, pp. 15–27, 1998.

29. V. López, S. Akerreta, E. Casanova, J. M. García-Mina, R. Y. Cavero, and M. I. Calvo, "In vitro antioxidant and anti-rhizopus activities of Lamiaceae herbal extracts," Plant Foods for Human Nutrition, vol. 62, no. 4, pp. 151–155, 2007.

30. P. P. Singh and P. Sharma, "Antioxidant basket: do not mix apples and oranges," Indian Journal of Clinical Biochemistry, vol. 24, no. 3, pp. 211–214, 2009.

31. M. Dong, X. He, and H. L. Rui, "Phytochemicals of black bean seed coats: isolation, structure elucidation, and their antiproliferative and antioxidative activities," Journal of Agricultural and Food Chemistry, vol. 55, no. 15, pp. 6044–6051, 2007.

32. C. Vijaya Kumar Reddy, D. Sreeramulu, and M. Raghunath, "Antioxidant activity of fresh and dry fruits commonly consumed in India," Food Research International, vol. 43, no. 1, pp. 285–288, 2010.

33. K. Mahattanatawee, J. A. Manthey, G. Luzio, S. T. Talcott, K. Goodner, and E. A. Baldwin, "Total antioxidant activity and fiber content of select Florida-grown tropical fruits," Journal of Agricultural and Food Chemistry, vol. 54, no. 19, pp. 7355–7363, 2006.

34. J. Sun, Y. Chu, X. Wu, and R. H. Liu, "Antioxidant and antiproliferative activities of common fruits," Journal of Agricultural and Food Chemistry, vol. 50, no. 25, pp. 7449–7454, 2002.

35. N. M. A. Hassimotto, M. I. Genovese, and F. M. Lajolo, "Antioxidant activity of dietary fruits, vegetables, and commercial frozen fruit pulps," Journal of Agricultural and Food Chemistry, vol. 53, no. 8, pp. 2928–2935, 2005.

36. C. Kevers, M. Falkowski, J. Tabart, J. Defraigne, J. Dommes, and J. Pincemail, "Evolution of antioxidant capacity during storage of selected fruits and vegetables," Journal of Agricultural and Food Chemistry, vol. 55, no. 21, pp. 8596–8603, 2007.

37. D. Sreeramulu and M. Raghunath, "Antioxidant activity and phenolic content of roots, tubers and vegetables commonly consumed in India," Food Research International, vol. 43, no. 4, pp. 1017–1020, 2010.

38. E. H. Jeffery, A. F. Brown, A. C. Kurilich et al., "Variation in content of bioactive components in broccoli," Journal of Food Composition and Analysis, vol. 16, no. 3, pp. 323–330, 2003.

39. M. P. Kähkönen, A. I. Hopia, H. J. Vuorela et al., "Antioxidant activity of plant extracts containing phenolic compounds," Journal of Agricultural and Food Chemistry, vol. 47, no. 10, pp. 3954–3962, 1999.

40. U. Imeh and S. Khokhar, "Distribution of conjugated and free phenols in fruits: antioxidant activity and cultivar variations," Journal of Agricultural and Food Chemistry, vol. 50, no. 22, pp. 6301–6306, 2002.

41. J. O. Kuti and H. B. Konuru, "Antioxidant capacity and phenolic content in leaf extracts of tree spinach (Cnidoscolus spp.)," Journal of Agricultural and Food Chemistry, vol. 52, no. 1, pp. 117–121, 2004.

42. A. L. K. Faller and E. Fialho, "The antioxidant capacity and polyphenol content of organic and conventional retail vegetables after domestic cooking," Food Research International, vol. 42, no. 1, pp. 210–215, 2009.

43. A. Bunea, M. Andjelkovic, C. Socaciu et al., "Total and individual carotenoids and phenolic acids content in fresh, refrigerated and processed spinach (Spinacia oleracea L.)," Food Chemistry, vol. 108, no. 2, pp. 649–656, 2008.

44. J. A. Tudela, E. Cantos, J. C. Espín, F. A. Tomás-Barberán, and M. I. Gil, "Induction of antioxidant flavonol biosynthesis in fresh-cut, potatoes. Effect of domestic cooking," Journal of Agricultural and Food Chemistry, vol. 50, no. 21, pp. 5925–5931, 2002.

45. A. M. Jimenez-Monreal, L. Garcia-Diz, M. Martinez-Tome, M. Mariscal, and M. A. Murcia, "Influence of cooking methods on antioxidant activity of vegetables," Journal of Food Science, vol. 74, no. 3, pp. H97–H103, 2009.

46. M. Racchi, M. Daglia, C. Lanni, A. Papetti, S. Govoni, and G. Gazzani, "Antiradical activity of water soluble components

in common diet vegetables," Journal of Agricultural and Food Chemistry, vol. 50, no. 5, pp. 1272–1277, 2002.

47. N. Turkmen, F. Sari, and Y. S. Velioglu, "The effect of cooking methods on total phenolics and antioxidant activity of selected green vegetables," Food Chemistry, vol. 93, no. 4, pp. 713–718, 2005.

48. B. Chipurura, Z. M. Muchuweti, and F. Manditseraa, "Effects of thermal treatment on the phenolic content and antioxidant activity of some vegetables," Asian Journal of Clinical Nutrition, vol. 2, no. 3, pp. 93–100, 2010.

49. M. Bajpai, A. Mishra, and D. Prakash, "Antioxidant and free radical scavenging activities of some leafy vegetables," International Journal of Food Sciences and Nutrition, vol. 56, no. 7, pp. 473–481, 2005.

Purslane Weed (Portulaca oleracea): A Prospective Plant Source of Nutrition, Omega-3 Fatty Acid, and Antioxidant Attributes

Md. Kamal Uddin[1,] Abdul Shukor Juraimi[1,] Md Sabir Hossain[1,] Most. Altaf Un Nahar[2,] Md. Eaqub Ali[3,] and M. M. Rahman[4]

[1]Institute of Tropical Agriculture, University Putra Malaysia, Serdang, Malaysia

[2]Bangladesh Agricultural University, Mymensingh 2202, Bangladesh

[3]Malaysia Nanotechnology and Catalysis Research Center, University of Malaya, 50603 Kuala Lumpur, Malaysia

[4]Bangabadhu Sheikh Mujibur Rahman Agricultural University, Gazipur 1706, Bangladesh

ABSTRACT

Purslane (Portulaca oleracea L.) is an important plant naturally found as a weed in field crops and lawns. Purslane is widely distributed around the globe and is popular as a potherb in many areas of Europe, Asia, and the Mediterranean region. This plant possesses mucilaginous substances which are of medicinal importance. It is a rich source of potassium (494 mg/100 g) followed by magnesium (68 mg/100 g) and calcium (65 mg/100 g) and possesses the potential to be used as vegetable source of omega-3 fatty acid. It is very good source of alpha-linolenic acid (ALA) and gamma-linolenic acid (LNA, 18:3 w3) (4 mg/g fresh weight) of any green leafy vegetable. It contained the highest amount (22.2 mg and 130 mg per 100 g of fresh and dry weight, resp.) of alpha-tocopherol and ascorbic acid (26.6 mg and 506 mg per 100 g of fresh and dry weight, resp.). The oxalate content of purslane leaves was reported as 671–869 mg/100 g fresh weight. The antioxidant content and nutritional value of purslane are important for human consumption. It revealed tremendous nutritional potential and has indicated the potential use of this herb for the future.

INTRODUCTION

Purslane (Portulaca oleracea L.) deserves special attention from agriculturalists as well as nutritionists. Purslane is a common weed in turfgrass areas as well as in field crops [1, 2]. Many varieties of purslane under many names grow in a wide range of climates and regions. Purslane has wide acceptability as a potherb in Central Europe, Asia, and the Mediterranean region. It is an important component of green salad and its soft stem and leaves are used raw, alone, or with other greens. Purslane is also used for cooking or used as a pickle. Its medicinal value is evident from its use for treatment of burns, headache, and diseases related to the intestine, liver, stomach, cough, shortness of breath, and arthritis. Its use as a purgative, cardiac tonic, emollient, muscle relaxant, and anti-inflammatory and diuretic treatment makes it important in herbal medicine. Purslane has also been used in the treatment of osteoporosis and psoriasis.

Recent research demonstrates that purslane has better nutritional quality than the major cultivated vegetables, with higher beta-carotene, ascorbic acid, and alpha-linolenic acid [3]. Additionally, purslane has been described as a power food because of its high nutritive and antioxidant properties [4]. Different varieties, harvesting times, and environmental conditions can contribute to purslane's nutritional composition and benefits [5].

Purslane is popular as a traditional medicine in China for the treatment of hypotension and diabetes. Scientifically, it is not proven to have antidiabetic effects, but still people use it for this purpose. An experiment has been carried out for the extraction of crude polysaccharide(s) from purslane to investigate the hypoglycemic effects of these constituents with animal tests for the use of this plant in the treatment of diabetes [6].

Purslane is a very good source of alpha-linolenic acid. Alpha-linolenic is an omega-3 fatty acid which plays an important role in human growth and development and in preventing diseases. Purslane has been shown to contain five times higher omega-3 fatty acids than spinach. Omega-3 fatty acids belong to a group of polyunsaturated fatty acids essential for human growth, development, prevention of numerous cardiovascular diseases, and maintenance of a healthy immune system [7]. Our bodies do not synthesise omega-3 fatty acids. Therefore omega-3 fatty acids must be consumed from a dietary source. Omega-3 fatty acids contain 18 to 24 carbon atoms and have three or more double bonds within its fatty acid chain [8]. Fish is the richest source of omega-3 fatty acids. Health authorities highly recommend that we consume fish regularly to meet our bodies' requirements of omega-3 fatty acids, as other sources are limited and do not supply nearly as much omega-3 fatty acids [9]. Purslane has recently been identified as the richest vegetable source of alpha-linolenic acid, an essential omega-3 fatty acid [10]. The lack of dietary sources of omega-3 fatty acids has resulted in a growing level of interest to introduce purslane as a new cultivated vegetable [11, 12]. Purslane flourishes in numerous biogeographical locations worldwide and is highly adaptable to many adverse conditions such as drought, saline, and nutrient deficient conditions [13].

Distribution: It is reported that purslane was a common vegetable of the Roman Empire. Origin of purslane is not certain, but existence of this plant is reported about 4,000 years ago. The succulent stems and fleshy leaves of purslane reflect that it may have originated and adapted to desert climates of the Middle East and India. It can be found in Europe, Africa, North America, Australia, and Asia [14].

Botanical Classification: Portulaca oleracea is s cosmopolitan species and the genus Portulaca belongs to the family Portulacaceae, a small family with 21 genera and 580 species, and is cosmopolitan in distribution, occurring especially in America with some species found in Arabia [15]. Purslane plants are succulent, annual herbaceous, and erect or decumbent up to 30 cm high. Purslane is botanically known as Portulaca oleracea and is also called portulaca.

Habitat: It grows well in orchards, vineyards, crop fields, landscaped areas, gardens, roadsides, and other disturbed sites.

Stem: Stems are cylindrical, up to 30 cm long, 2-3 mm in diameter, green or red, swollen at the nodes, smooth, glabrous apart from the leaf axils, and diffusely branched, and the internodes are 1.5–3.5 cm in length.

Leaf: Purslane leaves are alternate or subopposite, flat, fleshy, having variable shapes, obovate, 1–5 cm long, 0.5–2 cm across, obtuse or slightly notched at the apex, tapering at base, sessile or indistinctly petiolate, glabrous, smooth, and waxy on the upper surface, with entire margin, small stipules, and cluster of hairs up to 1 mm long. Leaves are egg to spatula shaped, succulent, and stalkless or have very short stalks, about 5–30 mm long, and sometimes their edges are red-tinged. Leaves are green or green with red margin.

Seedling: Cotyledons (seed leaves) are egg shaped to oblong, hairless, succulent, about 2–5 mm long, and sometimes tinged red.

Flower: Flowering initiates during May to September. Flowers originate as single or clusters of two to five at the tips of stems. The flowers are minute or small having orange yellow, purple, or white pink color with five petals and typically open only on hot and sunny days from mid-morning to early afternoon.

Fruit: Fruit consists of almost round to egg-shaped capsules, usually about 4–8 mm long that open around the middle to release the seeds. Seeds are tiny, less than 1 mm in diameter, circular to egg shaped, flattened, and brown to black with a white point of attachment. Numerous seeds are produced.

HEALTH BENEFITS OF PURSLANE

Nutrition

It is rich in vitamin A which is a natural antioxidant value. It can play role in vision healthy mucus membranes and to protect from lung and oral cavity cancer. Purslane contains the highest content of vitamin A among green leafy vegetables. It also contains vitamin C and B-complex vitamins like riboflavin, niacin, and pyridoxine. It provides highest dietary minerals such as potassium (494 mg/100 g) followed by magnesium (68 mg/100 g), calcium (65 mg/100 g), phosphorus (44 mg/100 g), and iron (1.99 mg/100 g) (Table 1).

Table 1: Purslane (Portulaca oleracea) (Nutritive value per 100 g)

Principle	Nutrient value	Percentage of RDA
Energy	16 Kcal	1.5%
Carbohydrates	3.4 g	3%
Protein	1.30 g	2%
Total Fat	0.1 g	0.5%
Cholesterol	0 mg	0%
Vitamins		
Folates	12 µg	3%
Niacin	0.480 mg	3%
Pantothenic acid	0.036 mg	1%
Pyridoxine	0.073 mg	5.5%
Riboflavin	0.112 mg	8.5%
Thiamin	0.047 mg	4%
Vitamin A	1320 IU	44%
Vitamin C	21 mg	35%
Electrolytes		
Sodium	45 mg	3%
Potassium	494 mg	10.5%
Minerals		
Calcium	65 mg	6.5%
Copper	0.113 mg	12.5%
Iron	1.99 mg	25%

Magnesium	68 mg	17%
Manganese	0.303 mg	13%
Phosphorus	44 mg	6%
Selenium	0.9 µg	2%
Zinc	0.17 mg	1.5%

Source: USDA National Nutrient data.

The range of Ca, Mg, K, Fe, and Zn from the young stage to mature plants was from 1612 ± 27 to 1945 ± 30 mmol kg^{-1} DW, 2127 ± 23 to 2443 ± 27 mmol kg^{-1} DW, 1257 ± 10 to 1526 ± 31 mmol kg^{-1} DW, 218 ± 8 to 262 ± 3 mmol kg^{-1} DW, and 128 ± 2 to 160 ± 1 mmol kg^{-1} DW, respectively. On the other hand, the Na and Cl concentrations in leaves were higher at the young stage and lower at the mature stage. The Na and Cl concentrations ranged from 356 ± 4 to 278 ± 8 mmol kg^{-1} DW and from 82 ± 2 to 53 ± 2 mmol kg^{-1} DW, respectively [16].

Omega-3 Fatty Acid

Purslane is one of the richest green plant sources of omega-3 fatty acids. It has lower the cholesterol and triglyceride levels, raise the beneficial high density lipoprotein. Moreover, the ability of omega-3 fatty acids to decrease the thickness of the blood may be advantageous in the treatment of vascular diseases [3]. Unlike fish oils with their high cholesterol and calorie content, purslane also provides an excellent source of the beneficial omega-3 fatty acids without the cholesterol of fish oils, since it contains no cholesterol. There are 3 varieties of purslane, namely, the green, golden, and a large-leaved golden variety [17, 18]. Important sources of omega-3 fatty acids are summarized in Table 2. It has a low incidence of cancer and heart disease, possibly due in part to purslane's naturally occurring omega-3 fatty acids [19].

Table 2: Plant sources of omega-3 fatty acids (g/100 g)

Category	Fruits/vegetables	Amount (g)
Low	Avocados, California raw	0.1
	Broccoli	0.1
	Strawberries	0.1
	Cauliflower, raw	0.1
	Kale, raw	0.2
	Spinach, raw	0.1
	Peas, garden dry	0.2
	Cowpeas, dry	0.3
	Beans, navy, sprouted, cooked	0.3
	Corn, germ	0.3
Medium	Bean, common dry	0.6
	Leeks, freeze-dried, raw	0.7
	Wheat, germ	0.7
	Spirulina, dried	0.8
	Purslane	0.9
	Oat, germ	1.4
	Beachnuts	1.7
	Soybeans kernels, roasted	1.5
	Soybeans, green	3.2
High	Soybean oil	6.8
	Walnuts, Persian, English	6.8
	Wheat germ oil	6.8
	Butternuts	8.7
	Walnut oil	10.4
	Rapeseed oil (New Puritan Oil)	11.1

Source: Bulletin, US Department of Agriculture Provisional table on the content of omega-3 fatty acids and other fat components in selected foods (HNIS/PT-103).

Purslane is best used for human consumption as a green vegetable rich in minerals and omega-3 fatty acids [20]. Omega-3 fatty acid is a precursor of a specific group of hormones. It may offer protection against cardiovascular disease, cancers, and a number of chronic diseases and conditions throughout the human life. The antioxidant enzymes such as GPx, GR, SOD, and GST take part in maintaining glutathione homeostasis in tissues. Also, increased levels of GPx, GR,

GST, CAT, and SOD were found to correlate with elevated glutathione level and depressed MDA and NO in rats, thus showing the antioxidant activity of purslane.

Purslane leaves contain higher contents of alpha-linolenic acid (18:3 w3), alpha-tocopherol, ascorbic acid and glutathione than the leaves of spinach. It grows in growth chambers containing seven times higher contents of alpha-tocopherol than that found in spinach. One hundred grams of fresh purslane leaves (one serving) contains about 300–400 mg of 18:3 w3; 12.2 mg of alpha-tocopherol, 26.6 mg of ascorbic acid, 1.9 mg of beta-carotene, and 14.8 mg of glutathione [21].

Purslane has the highest level of alpha-linolenic which is an omega 3 fatty acid essential for human nutrition compared to any leafy green vegetable. A 100 g sample of purslane contains 300–400 mg of alpha-linolenic acid (ALA). It also has 0.01 mg per gram of eicopentanoic acid (EPA), which is not present at all in flax oil. This would provide 1 mg of EPA for a 100 g portion of purslane or 10 mg for a kg (2.2 pounds), or 1 g for 100 kg (220 pounds) of sample.

Purslane is the richest source of gamma-linolenic acid (LNA, 18:3 w3) (4 mg/g fresh weight) of any green leafy vegetable. It also contains a small amount of eicosapentaenoic acid (EPA. 20:5 w3) (0.01 mg/g fresh weight) [21]. Subsequently, purslane contained 18:3 w3 20:5 w3 and 22:6 w3 (docosahexaenoic acid, DHA) as well as 22:5 w3 (docosapentaenoic acid, DPA) [22].

Selected food sources of omega-3 fatty acids are illustrated in Table 2. Rapeseed oil, walnut oil, butternuts, and wheat germ oil are excellent sources (6.8–11.1 g/100 g fresh weight) of omega-3 fatty acids. Good sources (0.6–3.2 g/100 g fresh weight) of these fats include green soybean, soybean kernels, beechnut, and oat germ. Cabbage, cauliflower, broccoli, strawberries, spinach, garden pea, corn, and common dry bean are additional sources of omega-3 fatty acids. These foods contain a smaller (0.1–0.3 g/100 g fresh weight) level of omega-3 fatty acids.

Purslane leaves contained higher amounts of alpha-linolenic (18:3 w3) than stem fractions, whereas 20:5w3 was higher in stem fractions [22] (Table 3).

Table 3: Composition of selected fatty acids in purslane (Portulaca oleracea) (% of total FA)[a]

Omara-Alwala et al., 1991 [22]				Simopoulos and Salem, 1986 [10]
Fatty acid	Leaf	Stem	Whole plant	Whole plant
18.3-omega-3	41.4–66.4	2.4–5.9	28.4–42.5	47.6
20.5-omega-3	0.8–12.6	18.6–35.5	6.4–21.5	0.1
22.3-omega-3	1.4–3.3	trace	1.0–3.0	—
22.6-omega-3	0.3–6.4	trace	0.6–5.6	—

Results from Omara-Alwala et al., 1991 [22], and Simopoulos and Salem, 1986 [10], expressed as mg of FA per kg or g of net weight.

Leaves of purslane grown both in the controlled growth chamber and in the wild contained higher amount of alpha-linolenic fatty acid (18: 3 w3) than that of spinach leaves. The highest amount (3.41 mg/g) of alpha-linolenic acid was recorded in growth chamber grown purslane, which was seven times higher than that of spinach leaves (0.48 mg/g) (Table 4).

Table 4: Fatty acid profiles in total lipid extracts from leaves of purslane and spinach

Fatty acid	Chamber grown purslane		Wild purslane		Spinach	
	Dry wt%	mg/g fresh wt	Dry wt%	mg/g fresh wt	Dry wt%	mg/g fresh wt
18.0	1.12	0.064	0.95	0.048	0.78	0.007
18.1	4.99	0.016	2.13	0.10	2.04	0.018
18.2	16.99	0.968	13.45	0.70	11.70	0.10
18.3	59.87	3.41	63.78	3.22	53.85	0.48

Source: Simopoulos et al., 1992 [21]

Lipid Content and Fatty Acid Composition

All fractions contained very low lipid content with 0.47% in stems, 0.51% in leaves, and 0.54% in the flowers (Table 5). In general, polyunsaturated fatty acids (PUFAs) were found to be most abundant in all fractions, followed by saturated (SFAs) and monounsaturated fatty acids (MUFAs). The most predominant fatty acids were 18:3n-3 (50%) in the leaf, 18:3n-6 (46%) in the stem, and 18:2n-6 (30% of total fatty acid) in the flowers. ALA content ranged from 149 to 523 mg (100 g sample) in stems and leaves, respectively. An interesting finding in this study was that 18:3n-6 was found at high levels in all fractions, accounting for 46% in stems, 13% in leaves, and 10% in flowers [23].

Table 5: Fatty acid composition of purslane fractions

Fatty acid	Composition (% of total fatty acids)		
	Leaf	Stem	Flower
15:0	0.39	—	1.01
16:0	13.09	16.90	19.30
18:0	2.29	7.75	4.51
Total SFA	16.42	24.64	24.52
16:1	0.54	0.51	0.90
18:1	4.29	3.38	12.30
Total MUFA	4.83	3.89	13.20
18:2n-6	14.46	9.70	30.11
18:3n-6	13.25	45.57	9.68
18:3n-3	49.70	15.62	21.01
20:0	0.21	0.11	0.29
22:0	0.19	0.16	0.10
24:0	1.12	0.31	1.09
Total PUFA	78.75	71.47	62.28
Lipid content (%)	0.51	0.47	0.54

Source: Siriamornpun and Suttajit, 2010 [23].

Antioxidants

The TPC in cultivars of P. oleracea ranged from 127 ± 13 to 478 ± 45 mg GAE/100 g fresh weight of plant. The IC50 ranged from 0.89 ± 0.07 to 3.41 ± 0.41 mg/mL, the AEAC values ranged from 110 ± 14 to 430 ± 32 mg AA/100 g, and the FRAP values ranged from 0.93 ± 0.22 to 5.10 ± 0.56 mg GAE/g [24] (Lim and Quah 2007). DPPH scavenging (IC50) capacity ranged from 1.30 ± 0.04 to 1.71 ± 0.04 mg/mL, while the ascorbic acid equivalent antioxidant activity (AEAC) values were from 229.5 ± 7.9 to 319.3 ± 8.7 mg AA/100 g, the total phenol content (TPC) varied from 174.5 ± 8.5 to 348.5 ± 7.9 mg GAE/100 g, AAC varied from 60.5 ± 2.1 to 86.5 ± 3.9 mg/100 g, and FRAP ranged from 1.8 ± 0.1 to 4.3 ± 0.1 mg GAE/g [16].

Higher amounts of alpha-tocopherol, ascorbic acid, and beta-carotene were observed in the leaves of purslane grown both in the growth chamber and in the wild, compared to the composition of spinach leaves (Table 6). Growth chamber grown purslane contained the highest amount (22.2 mg and 130 mg per 100 g of fresh and dry weight, resp.) of alpha-tocopherol and ascorbic acid (26.6 mg and 506 mg per 100 g of fresh and dry weight, resp.), whereas beta-carotene was slightly higher in spinach.

Table 6: Antioxidant content of purslane and spinach leaves

Name	Alpha-tocopherol		Ascorbic acid		Beta-carotene	
	Mg/100 g (fresh wt.)'	Mg/100 g (dry wt.)	Mg/100 g (fresh wt.)'	Mg/100 g (dry wt.)	Mg/100 (g fresh wt.)'	Mg/100 g (dry wt.)
Chamber grown purslane	22.2	230	26.6	506	1.9	38.2
Wild purslane	8.2	170	23.0	451	2.2	43.5
Spinach	1.8	36	21.7	430	3.3	63.5

Source: Simopoulos et al, 1992 [21].

Vitamin C (ascorbic acid) and beta-carotene have been reported to possess antioxidant activity, because of their ability to neutralize free radicals, and have the potential to prevent cardiovascular disease and cancer [25]. Leaves had the highest content of beta-carotene, ascorbic acid, and DPPH, followed by flowers and stems (Table 7). Thai wild purslane contained almost 10 time's higher beta-carotene and ascorbic acid [3, 4, and 26] content than other varieties. The beta-carotene content in the leaf was two times higher than in the stems and slightly higher than in the flowers. This finding is in agreement with the data on Australian purslane, where the beta-carotene content in the leaf was higher than in the stem [3]. Purslane is amongst the group of plants with high oxalate contents. Melatonin is a ubiquitous and versatile molecule that exhibits most of the desirable characteristics of a good antioxidant [27]. The oxalate content of purslane leaves was reported as 671–869 mg/100 g fresh weight [28, 29].

Table 7: Ascorbic acid and beta-carotene content and 1, 1-diphenyl-2-picryl-hydrazyl (DPPH) radical-scavenging activity of different parts of Thai purslane

Plant parts	Antioxidant compound content		
	Beta-carotene (mg g^{-1} sample)	Ascorbic acid (mg g^{-1} sample)	DPPH (%)
Leaf	0.58	3.99	76.71
Stem	0.29	2.27	90.11
Flower	0.55	2.32	91.01

CONCLUSIONS

As a significant source of omega-3 oils, P. oleracea could yield considerable health benefits to vegetarian and other diets where the consumption of fish oils is excluded. Scientific analysis of its chemical components has shown that this common weed has uncommon nutritional value, making it one of the potentially important foods for the future. Presence of high content of antioxidants (vitamins A and C, alpha-tocopherol, beta-carotene, and glutathione) and omega-3 fatty acids and its wound healing and antimicrobial effects as well as its traditional use in the topical treatment of inflammatory conditions suggest that purslane is a highly likely candidate as a useful cosmetic ingredient.

REFERENCES

1. M. D. Kamal-Uddin, A. S. Juraimi, M. Begum, M. R. Ismail, A. A. Rahim, and R. Othman, "Floristic composition of weed community in turf grass area of west peninsular Malaysia," International Journal of Agriculture and Biology, vol. 11, no. 1, pp. 13–20, 2009.

2. M. K. Uddin, A. S. Juraimi, M. R. Ismail, and J. T. Brosnan, "Characterizing weed populations in different turfgrass sites throughout the Klang Valley of western Peninsular Malaysia," Weed Technology, vol. 24, no. 2, pp. 173–181, 2010.

3. L. Liu, P. Howe, Y.-F. Zhou, Z.-Q. Xu, C. Hocart, and R. Zhang, "Fatty acids and β-carotene in Australian purslane (Portulaca oleracea) varieties," Journal of Chromatography A, vol. 893, no. 1, pp. 207–213, 2000.

4. A. P. Simopoulos, H. A. Norman, and J. E. Gillaspy, "Purslane in human nutrition and its potential for world agriculture," World Review of Nutrition and Dietetics, vol. 77, pp. 47–74, 1995.

5. I. Oliveira, P. Valentão, R. Lopes, P. B. Andrade, A. Bento, and J. A. Pereira, "Phytochemical characterization and radical scavenging activity of Portulaca oleraceae L. leaves and stems," Microchemical Journal, vol. 92, no. 2, pp. 129–134, 2009.

6. A. P. Simopoulos, "Omega-3 fatty acids and antioxidants in edible wild plants," Biological Research, vol. 37, no. 2, pp. 263–277, 2004.

7. I. Gill and R. Valivety, "Polyunsaturated fatty acids. Part 1: occurrence, biological activities and applications," Trends in Biotechnology, vol. 15, no. 10, pp. 401–409, 1997.

8. J. Whelan and C. Rust, "Innovative dietary sources of n-3 fatty acids," Annual Review of Nutrition, vol. 26, pp. 75–103, 2006.

9. P. J. Nestel, "Polyunsaturated fatty acids (n-3, n-6)," The American Journal of Clinical Nutrition, vol. 45, no. 5, pp. 1161–1167, 1987.

10. A. P. Simopoulos and N. Salem Jr., "Purslane: a terrestrial source of omega-3 fatty acids," The New England Journal of Medicine, vol. 315, no. 13, p. 833, 1986.

11. U. R. Palaniswamy, R. J. McAvoy, and B. B. Bible, "Stage of harvest and polyunsaturated essential fatty acid concentrations in purslane (Portulaca oleraceae) leaves," Journal of Agricultural and Food Chemistry, vol. 49, no. 7, pp. 3490–3493, 2001.

12. I. Yazici, I. Türkan, A. H. Sekmen, and T. Demiral, "Salinity tolerance of purslane (Portulaca oleraceae L.) is achieved by enhanced antioxidative system, lower level of lipid peroxidation and proline accumulation," Environmental and Experimental Botany, vol. 61, no. 1, pp. 49–57, 2007.

13. M. K. Uddin, A. S. Juraimi, M. A. Hossain, F. Anwar, and M. A. Alam, "Effect of salt stress of Portulaca oleracea on antioxidant properties and mineral compositions," Australian Journal Crop Science, vol. 6, pp. 1732–1736, 2012.

14. A. N. Rashed, F. U. Afifi, and A. M. Disi, "Simple evaluation of the wound healing activity of a crude extract of Portulaca oleracea L. (growing in Jordan) in Mus musculus JVI-1," Journal of Ethnopharmacology, vol. 88, no. 2-3, pp. 131–136, 2003.

15. A. Danin and J. A. Reyes-Betancort, "The status of Portulaca oleracae in the Canary Islands," Lagascalia, vol. 26, pp. 71–81, 2006.

16. M. K. Uddin, A. S. Juraimi, M. E. Ali, and M. R. Ismail, "Evaluation of antioxidant properties and mineral composition of Portulaca oleracea (L.) at different growth stages," International Journal Molecule Science, vol. 13, pp. 10257–10267, 2012.

17. A. P. Simopoulos, "Evolutionary aspects of diet, essential fatty acids and cardiovascular disease,"European Heart Journal, vol. 3, pp. D8–D21.

18. S. Siriamornpun and M. Suttajit, "Microchemical components and antioxidant activity of different morphological parts of thai wild purslane (Portulaca oleracea)," Weed Science, vol. 58, no. 3, pp. 182–188, 2010.

19. M. A. Dkhil, A. E. A. Moniem, S. Al-Quraishy, and R. A. Saleh, "Antioxidant effect of purslane (Portulaca oleracea) and its mechanism of action," Journal of Medicinal Plant Research, vol. 5, no. 9, pp. 1589–1593, 2011.

20. A. E. Abdel-Moneim, M. A. Dkhil, and S. Al-Quraishy, "The redox status in rats treated with flaxseed oil and lead-induced hepatotoxicity," Biological Trace Element Research, vol. 143, no. 1, pp. 457–467, 2011.

21. A. P. Simopoulos, H. A. Norman, J. E. Gillaspy, and J. A. Duke, "Common purslane: a source of omega-3 fatty acids and antioxidants," Journal of the American College of Nutrition, vol. 11, no. 4, pp. 374–382, 1992.

22. T. R. Omara-Alwala, T. Mebrahtu, D. E. Prior, and M. O. Ezekwe, "Omega-three fatty acids in purslane (Portulaca oleracea) Tissues," Journal of the American Oil Chemists› Society, vol. 68, no. 3, pp. 198–199, 1991.

23. S. Siriamornpun and M. Suttajit, "Microchemical components and antioxidant activity of different morphological parts of thai wild purslane (Portulaca oleracea)," Weed Science, vol. 58, no. 3, pp. 182–188, 2010.

24. Y. Y. Lim and E. P. L. Quah, "Antioxidant properties of different cultivars of Portulaca oleracea," Food Chemistry, vol. 103, no. 3, pp. 734–740, 2007.

25. V. A. Rifici and A. K. Khachadurian, "Dietary supplementation with vitamins C and E inhibits in vitro oxidation of lipoproteins," Journal of the American College of Nutrition, vol. 12, no. 6, pp. 631–637, 1993.

26. M. A. Alam, A. S. Juraimi, M. Y. Rafii et al., "Evaluation of antioxidant compounds, antioxidant activities and mineral composition of 13 collected purslane (Portulaca oleracea L.) accessions," BioMed Research International, vol. 2014, Article ID 296063, 10 pages, 2014.

27. A. Galano, D. X. Tan, and R. J. Reiter, "Melatonin as a natural ally against oxidative stress: a physicochemical examination," Journal of Pineal Research, vol. 51, no. 1, pp. 1–16, 2011.

28. A. P. Simopoulos, D.-X. Tan, L. C. Manchester, and R. J. Reiter, "Purslane: a plant source of omega-3 fatty acids and melatonin," Journal of Pineal Research, vol. 39, no. 3, pp. 331–332, 2005.

29. A. I. Mohamed and A. S. Hussein, "Chemical composition of purslane (Portulaca oleracea)," Plant Foods for Human Nutrition, vol. 45, no. 1, pp. 1–9, 1994.

A Study of Antioxidant Activity, Enzymatic Inhibition and in Vitro Toxicity of Selected Traditional Sudanese Plants with Anti-Diabetic Potential

Yasmin Hilmi[1], Muna F Abushama[1], Haidar Abdalgadir[2], Asaad Khalid[2], and Hassan Khalid[1]

[1]Khartoum College of Medical Sciences, Khartoum, Sudan
[2]Medicinal and Aromatic Plants Research Institute, National Centre for Research, Khartoum, Sudan

ABSTRACT

Background

Diabetes mellitus is a chronic metabolic disease with life-threatening complications. Despite the enormous progress in conventional medicine and pharmaceutical industry, herbal-based medicines are still a common practice for the treatment of diabetes. This study evaluated ethanolic and aqueous extracts of selected Sudanese plants that are traditionally used to treat diabetes.

Methods

Extraction was carried out according to method described by Sukhdev et. al. and the extracts were tested for their glycogen phosphorylase inhibition, Brine shrimp lethality and antioxidant activity using (DPPH) radical scavenging activity and iron chelating activity. Extracts prepared from the leaves of *Ambrosia maritima*, fruits of *Foeniculum vulgare* and *Ammi visnaga*, exudates of *Acacia Senegal*, and seeds of *Sesamum indicum* and *Nigella sativa*.

Results

Nigella sativa ethanolic extract showed no toxicity on Brine shrimp Lethality Test, while its aqueous extract was toxic. All other extracts were highly toxic and ethanolic extracts of *Foeniculum vulgare* exhibited the highest toxicity. All plant extracts with exception of *Acacia senegal* revealed significant antioxidant activity in DPPH free radical scavenging assay.

Conclusions

These results highly agree with the ethnobotanical uses of these plants as antidiabetic. This study endorses further studies on plants investigated, to determine their potential for type 2 diabetes management. Moreover isolation and identification of active compounds are highly recommended.

BACKGROUND

Diabetes mellitus (DM) is a disease with severe complications and major health/economic impacts. It is a leading cause of morbidity and mortality worldwide, with an estimated 346 million adults being affected in year 2011. WHO projects that diabetes death will increase by two- thirds between 2008 and 2030 [1]. The prevalence of diabetes for all age-groups worldwide was estimated to be 2.8% in 2000 and 4.4% in 2030 [2].

In Sudan diabetes is an increasingly important problem, being responsible for 10% of hospital admissions and mortality [3]. Recently, an increase in incidence of DM has been observed especially among urbanized population indicating that diabetes mellitus is emerging as an important health problem [4]. The results of a small-scale study carried out in 1996 indicated that diabetes population in Sudan is at around one million, 90% of them have type 2 diabetes. It also showed a prevalence of 3.4% of type 2 DM [5].

Nature is an extraordinary source of medicines. The use of traditional medicines and medicinal plants in most developing countries as therapeutic agents for the maintenance of good health has been widely observed. The World Health Organization estimated that 80% of the populations of developing countries rely on traditional medicines, mostly plant drugs, for their primary health care needs. Diabetes is an example of a disease that has been treated with plant medicines. Research conducted in the last few decades on plants used traditionally for treatment of diabetes has shown antidiabetic properties [6].

To date, more than 1200 flowering plants have been claimed to have antidiabetic properties. Among them, only one-third have been scientifically studied and documented in around 460 publications [7].

The Sudanese flora has a vast variety of medicinal plants, which are traditionally used for their antidiabetic property. However, careful assessment including sustainability of such herbs, seasonal variation in activity of phyto-constituents, metal contents of crude herbal anti-diabetic drugs, thorough toxicity study and cost effectiveness is required for their popularity. These efforts may justify the role of novel traditional medicinal plants having anti-diabetic potentials.

Herbal drugs are considered free from side effects than synthetic one. They are less toxic, relatively cheap and popular [8]. However cytotoxicity of these plants needs to be monitored.

Brine shrimp bioassays (BSLT) offer a quick, simple and cost-efficient way of testing the toxicity of plant extracts. It has been developed for screening, fractionation and monitoring of biologically active natural products [9, 10].

Oxidative stress has been implicated in the development of many pathophysiological conditions including diabetes. Oxidative stress takes place due to the disturbance of the balance between the formation of reactive oxygen species (ROS) and the defense provided by cellular antioxidants. Medicinal plants provide a natural source of antioxidants that have been used worldwide for treatment of many diseases [6].

The inhibition of glycogen phosphorylase has been used as one method for treating type 2 diabetes. Since glucose production in the liver has been shown to increase in type 2 diabetes patients, inhibiting the release of glucose from the liver's glycogen's supplies appears to be a valid approach [11, 12].

In the present study we assessed the antioxidant and hypoglycemic activities of ethanolic and aqueous extracts of *Ambrosia maritima, Ammi visnaga, Acacia senegal, Sesamum indicum, Nigella sativa, Foeniculum vulgare* of Sudanese origin. Screening of toxicity of these plants using brine shrimp lethality test, is also investigated.

METHODS

Plant Material

Plants collected from Khartoum local market, identified and authenticated by Dr. Haider Abdelgadir, Herbarium Curator. Herbarium material was deposited at The Medicinal & Aromatic Plants Research Institute (MAPRI), Khartoum, Sudan. Table 1 shows the tested plants, their parts used and medicinal uses.

Table 1: Tested plants

Plant name	Family	Parts used	Reported medicinal uses[17-20]
Acacia senegal	Fabaceae–Mimosoidae	Fruits	Treatment of diabetes and chronic renal failure. Stem exudates (gums) are used as demulcents and against diarrhoea and ulcers
Ambrosia maritima	Asteraceae	Leaves	Hepatoprotective and Molluscicidal, The herbs are used in treatment of urinary tract infections and elimination of kidneystones, whereas the leaves are used as anti-diabetic and anti-hypertensive
Ammi visnaga	Apiaceae	Fruits	Used for renal urethra stones, smooth muscle relaxant
Foeniculum vulgare	Apiaceae	Fruits	Carminative, flatulence and digestive. It is also used in veterinary medicine
Nigella sativa	Ranunculaceae	Seeds	Treatment of diabetes, hypertension, abdominal ulcers, prostate gland and lung injury(Cuneyt Tayman)
Sesamum indicum	Pedaliaceae	Seeds	Treatment, for cough and cold. also used as nutritive, laxative, demulcent and emollient propertie

Hilmi *et al.*

Hilmi *et al. BMC Complementary and Alternative Medicine* 2014 14:149, doi:10.1186/1472-6882-14-149

Preparation of Plant Extracts

Extraction was carried out according to method described by Sukhdev *et. al.*[13].

Preparation of Ethanolic Extract

Specific weight of each plant sample was extracted by soaking in 96% ethanol for about seventy two hours with daily filtration and evaporation. Solvent was evaporated under reduced pressure to dryness using rotary evaporator apparatus and the extract combined together. The yield percentage was calculated as followed:

Weight of extract/weight of sample×100

Preparation of the Aqueous Extract

About 50–100 g of each plant sample was soaked in 500 ml hot distilled water, and left till cooled down with continuous stirring at room temperature. Extract was then filtered and freeze dried. Yield percentage was calculated.

Weight of extract obtained/weight of plantsample×100

Brine Shrimp Lethality Test

Artemia salina (shrimp eggs) was placed in natural sea water, and eggs hatched within 48 hrs, providing a large number of larvae (nauplii). The tested sample (20 mg) was dissolved in 2 ml of ethanol. From this solution 5, 50 and 500 µl were transferred to vials (triplicate for each concentration), forming concentrations of 10, 100 and 1000 µg/ml respectively. The solvent was allowed to evaporate overnight. Volume was made to 5 ml with seawater. 10 larvae were placed in each vial using a Pasteur pipette. Vials were incubated at 25–27°C for 24 hrs under illumination. Etoposide (7.4625 µg/ml) was used as positive control, and number of survived larvae were counted. Data was analyzed by Finney Probit Analysis computer program to determine LC_{50} values with 95% confidence intervals [9].

Antioxidant Activity Assays

DPPH Radical Scavenging Assay

The DPPH radical scavenging was determined according to the modified method of Shimada et al.[14] in 96-wells plate, the test samples were allowed to react with 1,1-Diphenyl-2-picryl-hydrazyl stable free radical (DPPH) for half an hour at 37°C. The concentration of DPPH was kept as 300 µM. The test samples were dissolved in DMSO while DPPH was prepared in ethanol. After incubation, decrease in absorbance was measured at 517 nm using multiple reader spectrophotometer. Percentage radical scavenging activity by samples was determined in comparison with a DMSO treated control group. All tests and analysis were run in triplicate. Vitamin C was used as positive control

Iron Chelating Activity Assay

The iron chelating ability was determined according to the modified method of Dinis et al.[15]. The Fe^{+2} were monitored by measuring the formation of ferrous ion-ferrozine complex. The experiment was carried out in 96 microtiter plate. The plant extracts were mixed with $FeSO_4$. The reaction was initiated by adding 5 mM ferrozine. The mixture was shaken and left at room temperature for 10 min. The absorbance was measured at 562 nm. EDTA was used as standard, and DMSO as control. All tests and analysis were run in triplicate.

Glycogen Phosphorylase Enzyme Assays

Glycogen phosphorylase a (from rabbit muscle), glycogen, glucose-1-phosphate, malachite green, and ammonium molybdate were purchased from the Sigma–Aldrich Corporation. Reagents and solvents were obtained from commercial suppliers and used without further purification. Solvents used were AR grade. The enzymatic inhibition of phosphorylase activity was monitored using multiskan spectrum (Thermo-Scientific) based on the published methods. In brief, GPa activity was measured in the direction of glycogen synthesis by the release of phosphate from glucose-1-phosphate. Each compound was dissolved in DMSO and diluted at different concentrations for IC50 determination. The enzyme was added into the 100 µL buffer with compounds dissolved in containing 50 mM Hepes (pH 7.2), 100 mM KCl, 2.5 mM MgCl2, 0.5 mM glucose-1-phosphate, and 1 mg/mL glycogen in 96-well microplates (costar). After the addition of 150 µL of 1 M HCl containing 10 mg/mL ammonium molybdate and 0.38 mg/mL malachite green, reactions were run at 25°C for 20 min. And then the phosphate absorbance was measured at 620 nm [16].

RESULTS AND DISCUSSION

Different medicinal systems that have been discovered as natural hypoglycemic medicine came from the virtue of traditional knowledge and have been used in many countries [7, 21, 22].

Many herbal extracts are currently traditionally used in Sudan for the treatment of diabetes. However, such medicinal plants have not gained much importance as medicines due to the lack of sustained scientific evidence. In the present study, aqueous and ethanolic extracts of six indigenous antidiabetic medicinal plants from Sudan were studied. The rationale for performing extractions from polar to non-polar solvents is to confirm and validate the inhibitory activity in the aqueous extractions performed in the traditional manner as well as to search for newer, more potent inhibitory compounds in the organic solvents. Since the Sudanese population has long been using all these plants for food and medicinal purposes, they form a part of the local pharmacopoeia.

Increasing evidence in both experimental and clinical studies suggests that oxidative stress plays a major role in the pathogenesis of both types of DM. Free radicals are formed disproportionately in diabetes by glucose oxidation, nonenzymatic glycation of proteins, and the subsequent oxidative degradation of glycated proteins. Abnormally high levels of free radicals and the simultaneous decline of antioxidant defense mechanisms can lead to damage of cellular organelles and enzymes, increased lipid peroxidation, and development of insulin resistance. These consequences of oxidative stress can promote the development of complications of DM. Changes in oxidative stress biomarkers, and use of antioxidants is an important trend in the treatment of DM [23].

In the present work we studied the anti-oxidant activity of ethanolic and aqueous extracts of investigated plants. Samples will be considered to have high or significant antioxidant capacity with IC50 < 50 µg/ml (extract) or IC50 < 10 µg/ml (compounds), moderate antioxidant capacity with 50 < IC50 < 100 µg/ml (extract) or 10 < IC50 < 20 µg/ml (compounds) and low antioxidant capacity with IC50 > 100 µg/ml (extract) or IC50 > 20 µg/ml (compounds) [24]. *Ambrosia maritima*, *Ammi visnaga* and *Foeniculum vulgare* exhibited high antioxidant activity in DPPH free radical scavenging assay with IC50 of 36 µg/ ml of ethanol extract of *Ambrosia maritima*, 41 µg/ ml of ethanol extract of *Ammi visnaga*, 47 µg/ml of aqueous extract of *Ammi visnaga*, 49 µg/ml of ethanol extract of *Foeniculum vulgare* and 31 µg/ml of aqueous extract of *Foeniculum vulgare*. This may support the traditional usage of these plants to improve complications such oxidative stress that caused by DM as well as many other diseases. A study in Egypt by Abu Zid and coworkers showed a moderate antioxidant activity of aqueous extract of *M. ambrosia* [25]. Results of *Achyranthes aspera* extracts in alloxan-treated mice revealed significant anti-hyperglycemic activity that may be mediated by diminished oxidative stress [26]. A study of *Eucalyptus globulus*, *Salvia officinalis* growing in Algeria and *Guiera senegalensis* growing in Sudan demonstrated that the 96% alcoholic leaf extracts had a significant blood-glucose lowering potential in glucose loaded rats with minimum toxicity [27]. Swanston and coworkers reported that agrimony, alfalfa, coriander, eucalyptus and juniper, can retard the development of streptozotocin diabetes in mice [28]. In another study, ethanolic crude extract of *Sorbus decora* demonstrates both anti-hyperglycemic and insulin-sensitizing activity *in vivo*, thereby confirming anti-diabetic potential and validating traditional medicine [29]. *Trigonella foenum-graecum*, *Atriplex halimus*, *Olea europaea*, *Urtica dioica*, *Allium sativum*, *Allium cepa*,*Nigella sativa*,and *Cinnamomum cassia* were tested for their antidiabetic properties. Results indicated that the observed anti-diabetic properties of these plants are mediated, at least partially, through regulating GLUT4 translocation [30].

Glycogen phosphorylase inhibition has been used as one method for treating type 2 diabetes[11,12]. Results of the current study did not show any significant inhibition of glycogen phosphorylase, but extracts of these plants may act on one of other enzymatic reactions that are involved in carbohydrate metabolism and improved glucose homeostasis.

All aqueous extracts showed significantly high toxicity on Brine shrimp Lethality Test, while *Foeniculum vulgare* showed moderate toxicity. Ethanolic extract of *Nigella sativa* showed no toxicity while all other ethanolic extracts exhibited high toxicity. Ethanolic extracts of *Foeniculum vulgare* exhibited the highest toxicity. These statistical consideration are based on the published work by Bussmann and coworkers. They stated that LC_{50} values below 249 µg/ml are considered as highly toxic, 250–499 µg/ml as median toxicity and 500–1000 µg/ml as light toxicity. Values above 1000 µg/ml are regarded as non-toxic [31]. These results could be very useful as preliminary data in the search for new antitumor compounds from the Sudanese market flora. All results for antioxidant activities, glycogen phosphorylase inhibition and cytotoxicity are shown in Table 2.

Table 2: Antioxidant activity, enzymatic inhibition and cytotoxicity of selected Sudanese medicinal plants

Plant	Extract	DPPH radical scavenging assay %	Iron chelating assay %	Inhibition % of glycogen phosphorylase (5mg/ml)	Brine shrimp lethality (LC 50)
Acacia Senegal	Ethanolic	Not Active	Not Active	0	83.8716
	Aqueous	Not Active	Not Active	0	17.9948
Ambrosia maritima	Ethanolic	60.8 ± 0.04	Not Active	2.2	39.7866
	Aqueous	21.2 ± 0.02	Not Active	0	10.6353
Ammi visnaga	Ethanolic	52.4 ± 0.03	Not Active	0	8.1217
	Aqueous	52.4 ± 0.03	2.5 ± 0.03	0	32.6273
Foeniculum vulgare	Ethanolic	60.7 ± 0.06	3.6 ± 0.05	0	0.012
	Aqueous	69.4 ± 0.003	Not Active	0	893.97
Nigella sativa	Ethanolic	47 ± 0.02	6.3 ± 0.02	0	11684.6
	Aqueous	19.3 ± 0.01	43.5 ± 0.04	0	122.268
Sesamum indicum	Ethanolic	Not Active	Not Active	8.2	61.85
	Aqueous	40.3 ± 0.01	23.2 ± 0.02	0	1.7

Hilmi *et al.*

Hilmi *et al. BMC Complementary and Alternative Medicine* 2014 14:149, doi: 10.1186/1472-6882-14-149

CONCLUSIONS

In conclusion these results revealed the significant antioxidant activity of the investigated plants extracts and may explain their role in altering the oxidative stress and management of diabetes mellitus. Furthermore the high toxicity of many extracts tested in this study suggests their antitumor potential and provides an avenue to explore the bioactive components of plant extracts. Studies should be directed towards drug industry by identification of single chemical compounds, and dosage use has to be monitored.

AUTHORS' CONTRIBUTIONS

YH participated in the study design and coordination, carried out the toxicity assay, drafted the manuscript and rewrote the final one. MA participated in the design of the study, toxicity assay and helped to draft the manuscript. HA conceived of the study, and participated in its design and coordination. AK participated in the enzymatic inhibition study and antioxidant activity. HK supervised part of the study and reviewed the manuscript. All authors read and approved the final manuscript.

ACKNOWLEDGEMENTS

The authors are grateful to the Medicinal & Aromatic Plants Research Centre (Khartoum, Sudan) for providing necessary laboratory facilities. We thank Mr. Muddathir Alhassan for his technical contribution in extraction methods. We thank Ms. Fatima Elfatih for her technical assistance in the enzymatic inhibition assays and Mr. Eltayeb Fadul for carrying out the IC50 analysis.

REFERENCES

1. *WHO Diabetes Fact Sheet N 312 September (2012)*. http://www. who.int/mediacentre/factsheets/fs312/en/index.html

2. Wild S, Bchir M, Roglic G, Green A, Sicree R, King H: Global Prevalence of Diabetes Estimates for the year 2000 and projections for 2030. *Diabetes Care* 2004, 27:1047-1053.

3. Ahmed AM, Ahmed NH, Abdalla ME: Pattern of hospital mortality among diabetic patients in Sudan. *Pract Diabetes Int* 2000, 17:41-43.

4. Ahmed AM: Diabetes mellitus in Sudan: size of the problem and possibilities of efficient care. *Pract Diabetes Int* 2001, 18:324-327.

5. Elbagir MN, Eltom MA, Elmahadi EM, Kadam IM, Berne C: A population-based study of the prevalence of diabetes and impaired glucose tolerance in adults in northern Sudan. *Diabetes Care* 1996, 19(10):1126-1128.

6. Abou El-Soud N, El-Laithy N, El-Saeed G, Wahby M, Khalil M, Morsy F, Shaffie N:Antidiabetic Activities of *Foeniculum Vulgare* Mill. Essential Oil in Streptozotocin-Induced Diabetic Rats. *Macedonian J Med Sci* 2011, 4(2):139-146.

7. Chang CL, Lin Y, Bartolome AB, Chen YC, Chiu SC, Yang WC: Herbal Therapies for Type 2 Diabetes Mellitus: Chemistry, Biology, and Potential Application of Selected Plants and Compounds. *Evidence-Based Complement Altern Med* 2013, 33.

8. Gupta R, Bajpai KG, Johri S, Saxena AM: An overview of Indian novel traditional medicinal plants with antidiabetic potentials. *Afr J Trad CAM* 2008, 5(1):1-17.

9. McLaughlin JL, Rogers LL, Anderson JE: The Use of Biological Assays to Evaluate Botanicals. *Drug Inf J* 1998, 32:513-524.

10. Bastos MLA, Lima MRF, Conserva LM, Andrade VS, Rocha EMM, Lemos RPL: Studies on the Antimicrobial Activity and Brine Shrimp Toxicity of Zeyheria tuberculosa (Vell.) Bur. (Bignoniceace) Extracts and Their Main Constituents. *Ann Clin Microbiol Antimicrob* 2009, 8:16-20.

11.   Somsák L, Nagya V, Hadady Z, Docsa T, Gergely P: Glucose analog inhibitors of glycogen phosphorylases as potential antidiabetic agents: recent developments. *Current Pharmacological Design* 2003, 9(15):1177-1189.

12.   Moller DE: New drug targets for type 2 diabetes and the metabolic syndrome. *Nature* 2001, 414(6865):821-827.

13.   Sukhdev SH, Suman PSK, Gennaro L, Dev DR: Extraction technologies for medicinal and aromatic plants. 2008. [*Chapter 1. United Nations Industrial Development Organization and the International Centre for Science and High Technology, Italy*]

14.   Shimada K, Fujikawa K, Yahara K, Nakamura T: Antioxidative properties of xanthan on the antioxidation of soybean oil in cyclodextrin emulsion. *J Agric Food Chem* 1992, 40(6):945-948.

15.   Dinis TCP, Madeira VMC, Almeida LM: Action of phenolic derivates (acetoaminophen, salycilate and 5-aminosalycilate) as inhibitor of membrane lipid peroxidation and as peroxyl radical scavengers. *Arch Biochem Biophys* 1994, 315:161-169.

16.   Martin WH, Hoover DJ: Discovery of a human liver glycogen phosphorylase inhibitor that lowers blood glucose in vivo. *Proc Natl Acad Sci U S A* 1998, 95(4):1776-1781.

17.   Khalid H, Abdella W, Abdelgadir H, Opatz T, Efferth T: Gems from traditional north-African medicine: medicinal and aromatic plants from Sudan Nat. *Prod Bioprospect* 2012, 2:92-103.

18.   Vanachayangkul P, Chow N, Khan SR, Butterweck V: Prevention of renal crystal deposition by an extract of *Ammi visnaga* L. and its constituent's khellin and visnagin in hyperoxaluric rats. *Urol Res* 2011, 39(3):189-195.

19.   Ghanem MTM, Radwan HMA, Mahdy EM, Elkholy YM, Hassanein HD, Shahat AA: Phenolic compounds from *Foeniculum vulgare* (Subsp. *Piperitum*) (Apiaceae) herb and evaluation of hepatoprotective antioxidant activity. *Pharmacognosy Res* 2012, 4(2):104-108.

20.   Tayman C, Cekmez F, Kafa I, Canpolat F, Cetinkaya M, Tonbul A, Uysal S, Tunc T, Sarici S: Protective Effects of Nigella sativa Oil in Hyperoxia-Induced Lung Injury. *Arch Bronconeumol* 2013, 49(1):15-21.

21. Baldé NM, Youla A, Baldé MD, Kaké A, Diallo MM, Baldé MA, Maugendre D: Herbal medicine and treatment of diabetes in Africa : an example from Guinea. *Diabetes Metab* 2006, 32(2):171-175.

22. Singh S, Gupta S, Sabir G, Gupta M, Seth P: Database for anti-diabetic plants with clinical/experimental trials. *Bioinformation* 2009, 4(6):263-268.

23. Maritim AC, Sanders RA, Watkins JB 3rd: Diabetes, oxidative stress, and antioxidants: a review. *J Biochem Mol Toxicol* 2003, 17(1):24-38.

24. Kuete V, Efferth T: Cameroonian medicinal plants: pharmacology and derived natural products. *Front Pharmacol* 2010, 1:123.

25. AbouZid S, Elshahaat A, Ali S, Choudhary M: Antioxidant activity of wild plants collected in Beni-Sueif. Upper Egypt. *Drug Discov Ther* 2008, 2(5):286-288. PubMed Abstract

26. Talukder F, Khan K, Uddin R, Jahan N, Alam A: In vitro free radical scavenging and anti-hyperglycemic activities of Achyranthes aspera extract in alloxan-induced diabetic mice. *Drug Discoveries Therapeut* 2012, 6(6):298-305.

27. Houacine CH, Elkhawad AO, Ayoub SMH: A comparative study on the anti-diabetic activity of extracts of some Algerian and Sudanese plants. *J Diabet Endocrin* 2012, 3(3):25-28.

28. Swanston-Flatt SK, Day C, Bailey CJ, Flatt PR: Traditional plant treatments for diabetes. Studies in normal and streptozotocin diabetic mice. *Diabetologia* 1990, 33(8):462-464.

29. Vianna R, Brault A, Martineau LC, Couture R, Arnason JT, Haddad PS: In Vivo Anti-Diabetic Activity of the Ethanolic Crude Extract of *Sorbus decora* C.K.Schneid. (Rosacea): A Medicinal Plant Used by Canadian James Bay Cree Nations to Treat Symptoms Related to Diabetes. *Evidence-Based Complement Altern Med* 2011, 7.

30. Kadan S, Saad B, Sasson Y, Zaid H: *In Vitro* Evaluations of Cytotoxicity of Eight antidiabetic Medicinal Plants and Their Effect on GLUT4 Translocation. *Evidence-Based Complement Altern Med* 2013, 9.

31. Bussmann R, Malca G, Glenn A, Sharon D, Nilsen B, Parris B, Dubose D, Ruiz D, Saleda A, Martinez M, Carillo L, Walker K,

Kuhlman A, Townesmith A: Toxicity of medicinal plants used in traditional medicine in Northern Peru. *J Ethnopharmacol* 2011, 137(1):121.

Chapter 7

Antioxidant Activity of Herbaceous Plant Extracts Protect against Hydrogen Peroxide-induced DNA Damage in Human Lymphocytes

Kuan-Hung Lin[1], Yan-Yin Yang[2], Chi-Ming Yang[3], Meng-Yuan Huang[3], Hsiao-Feng Lo[4], Kuang-Chuan Liu[5], Hwei-Shen Lin[2], and Pi-Yu Chao[6]

[1]Graduate Institute of Biotechnology, Chinese Culture University, Taipei 11114, Taiwan

[2]Graduate Institute of Applied Science of Living, Chinese Culture University, Taipei 11114, Taiwan

[3]Research Center for Biodiversity, Academia Sinica, Nankang, Taipei 11106, Taiwan

[4]Department of Horticulture and Landscape Architecture, National Taiwan University, Taipei 11111, Taiwan

[5]Taoyuan District Agricultural Research and Extension Station, Taoyuan 327 Taiwan

[6]Department of Nutrition and Health Sciences, Chinese Culture University, Taipei 11114, Taiwan

ABSTRACT

Background

Herbaceous plants containing antioxidants can protect against DNA damage. The purpose of this study was to evaluate the antioxidant substances, antioxidant activity, and protection of DNA from oxidative damage in human lymphocytes induced by hydrogen peroxide (H_2O_2). Our methods used acidic methanol and water extractions from six herbaceous plants, including *Bidens alba*(BA), *Lycium chinense* (LC), *Mentha arvensis* (MA), *Plantago asiatica* (PA), *Houttuynia cordata*(HC), and *Centella asiatica* (CA).

Methods

Antioxidant compounds such as flavonol and polyphenol were analyzed. Antioxidant activity was determined by the inhibition percentage of conjugated diene formation in a linoleic acid emulsion system and by trolox-equivalent antioxidant capacity (TEAC) assay. Their antioxidative capacities for protecting human lymphocyte DNA from H_2O_2-induced strand breaks was evaluated by comet assay.

Results

The studied plants were found to be rich in flavonols, especially myricetin in BA, morin in MA, quercetin in HC, and kaemperol in CA. In addition, polyphenol abounded in BA and CA. The best conjugated diene formation inhibition percentage was found in the acidic methanolic extract of PA. Regarding TEAC, the best antioxidant activity was generated from the acidic methanolic extract of HC. Water and acidic methanolic extracts of MA and HC both had better inhibition percentages of tail DNA% and tail moment as compared to the rest of the tested extracts, and significantly suppressed oxidative damage to lymphocyte DNA.

Conclusion

Quercetin and morin are important for preventing peroxidation and oxidative damage to DNA, and the leaves of MA and HC extracts may have excellent potential as functional ingredients representing potential sources of natural antioxidants.

BACKGROUND

Herbaceous plants have a long history of use as medicine, food, and a variety of daily needs. Many epidemiological studies suggest that an increased consumption of several medicinal plants containing antioxidants can protect against DNA damage and carcinogenesis, and often exhibit a wide range of pharmacological activities such as antiflammatory, anti-bacterial, and anti-fungal properties [1]. Flavonoids have strong antioxidant efficiencies and are common in leafy vegetables. Trolox, for example, is a water-soluble derivative of vitamin E that blocks DNA fragmentation in irradiated MOLT-4 cells, a human lymphocytic leukemia line [2]. Hence, a number of phytochemicals commonly used in research have antioxidant activity that can protect cells from reactive oxygen species (ROS)-mediated DNA damage that results in mutation and subsequent carcinogenesis [3,4]. Cao *et al.* [5] indicated that increased consumption of vegetables and fruits increases the plasma antioxidant capacity in humans. Some common vegetables like purple-leaved sweet potato and the outer layers of purple onions abound in quercetin and myricetin, which scavenge 2, 2-diphenyl-1-picrylhydrazyl (DPPH), superoxide, and hydroxyl radicals, and inhibit lipid peroxidation [6]. The search for phytochemicals and dietary compounds with potent antioxidant and otherwise preventive properties continues to be of great importance in the search for remedies against free radical-mediated diseases. There is great interest in the use of potent dietary antioxidants in preventive strategies for applications ranging from the prevention of oxidative reactions in foods and pharmaceuticals to the role of ROS in chronic degenerative diseases [7].

In recent years, increasing attention has been paid by consumers to the health and nutritional benefits of herbaceous plants. Some herbs, such as pilosa beggarticks (*Bidens alba* L. var. *minor*) (BA),

Chinese wolfberry (*Lycium chinense* Mill.) (LC), wild mint or corn mint (*Mentha arvensis* L. var. piperascens Malinv.) (MA), Asiatic plantain (*Plantago asiatica* L.) (PA), heartleaf (*Houttuynia cordata* Thunb.) (HC), and Asiatic centella (*Centella asiatica* L. Urban) (CA) are favored as functional herbals. Some of the health effects of herbaceous plants have been reported to include antioxidation [8-10], anti-inflammation [11], and blood pressure reduction [12]. In animal experiments, Chinese wolfberry, heartleaf, Asiatic plantain, Asiatic centella, and pilosa beggarticks showed special detoxification and anti-inflammatory effects [8, 9, 11, 13, 14]. Particularly, HC, LC, and CA showed antioxidant activities [8, 9]. Asiatic centella increased the activity of antioxidant enzymes such as superoxide dismutase, catalase, and glutathione peroxidase, and enhanced the concentration of vitamin C and vitamin E in new tissues during wound healings [13]. Both HC and BA were reported to have anti-inflammatory functions due to their quercetin and luteolin content [8, 11]. Furthermore, LC and BA can reduce the injury to liver cells from CCl_4 [9,13]. Pilosa beggarticks also functions as an anti-fungal and anti-bacterial agent, and lowers high blood pressure [12]. Several herbs are consumed to protect against common, serious diseases such as cardiovascular and cerebrovascular events, cancer, and other age-related degenerative diseases[15]. These protective effects are considered, in large part, to be related to the various antioxidants contained in them. Evidence that free radicals cause oxidative damage to lipids, proteins, and nucleic acids is overwhelming. Antioxidants, which can inhibit or delay the oxidation of an oxidizer in a chain reaction, would therefore seem to be important in preventing these diseases [16]. Prevention from oxidative stress might be achieved by the uptake of antioxidants. Polyphenols and flavonols can act as antioxidants in two ways: by scavenging free radicals and chelating redox active metal ions (direct antioxidant activity), and by inducing cellular antioxidant defense and repair. These benefits have significantly contributed to their antioxidant activity and have stimulated research into the content, ability, capacity, and function of antioxidant systems in herbaceous plants. Polyphenolic and flavonol substances are the most common compounds in herbs having strong antioxidant activity [6]. Previously, we also demonstrated that purple-leaved sweet potato exhibits free radical scavenging and has high polyphenolic content [17]. Although a variety of medicinal herbs are known to be potent sources of polyphenolic and flavonol compounds, studies that isolate polyphenols, evaluate their antioxidative effects, and determine their efficacy or ability to prevent oxidative damage to DNA are either scarce or little known. The bioactive components of these herbal plants might be responsible for anti-cancer effects through growth inhibition and apoptosis in human chronic myeloid leukemia K562 cells [18]. The objective of this study was to isolate, identify, and evaluate the antioxidant components, antioxidant activity, and extent to which methanolic acid hydrolysates and water extracts of six herbaceous plants could protect DNA in human lymphocytes from oxidative damage induced by H_2O_2. Our study explores the relationships between the composition and content of flavonols and polyphenol having antioxidant efficiency, and the prevention of DNA oxidative damage afforded by the herbaceous plants.

METHODS

Chemicals and Reagent

Methanol, ethanol, hydrochloric acid, di-sodium hydrogen phosphate, potassium dihydrogen phosphate, formic acid, sodium chloride (NaCl), potassium chloride (KCl), Tris–HCl, Tris (hydroxymethyl) aminomethane (Tris base), dimethyl sulfoxide (DMSO), ethylenediamine tetraacetic acid (EDTA), Trolox, and butylated hydroxyltoluene were purchased from Merck (Darmstadt, Germany). Linoleic acid, d-glucose, calcium chloride dihydrate, sodium lauryl sarcosinate, gallic acid, 2,2-azino-bis-(3-ethylbenzothiazoline-6-sulfonicacid) (ABTS), peroxidase, H_2O_2, sodium carbonate (Na_2CO_3), tetrazolium/formazan, Folin-Ciocalteau reagent, and ethidium bromide were procured from Sigma Chemical (St Louis, MO, USA). Myricetin, morin, quercetin, kaempferol, cynidin, and malvidin were obtained from ROTH (Rheinzabern, Denmark). Ficoll-Paque was acquired from Amersham Biosciences (Uppsala, Sweden). Low-melting gel agrose and Triton X-100 were purchased from BDH (Poole, England). Normal-melting gel agarose was purchased from Pantech Instruments (Darmstadt, Germany). AIM V serum-free lymphocyte medium was purchased from Gibco Invitrogen (Carlsbad, CA, USA).

Herbaceous Plants

The tested plants were *Bidens alba* L. var. *minor, Lycium chinense* Mill., *Mentha arvensis* L. var. piperascens Malinv., *Plantago asiatica* L., *Houttuyni acordata* Thunb., and *Centella asiatica* L. Urban. These were generously provided by Dr. Kuang-Chuan Liu, Taoyuan District Agricultural Research and Extension Station Council of Agriculture, Executive Yuan, Taiwan.

Preparation of Plant Extracts

The plants were weighed, lyophilized, and ground to powder. Each lyophilized powder was extracted by distilled deionized (dd) H_2O. The extraction mixture was then heated to 90°C in a steam bath and refluxed for 2 h, allowed to cool in a refrigerator, sonicated for 5 min, and diluted to 50 mL with ddH_2O to prepare the final extract. These water extracts were ready for the comet assay. For high-performance liquid chromatography (HPLC), only the edible portions of plants were weighed, lyophilized, and ground into powder. Lyophilized vegetable powders were prepared according to Justesen *et al.* [19] with modifications as follows: 10 ml of 62.5% aqueous methanol containing butylated hydroxyltoluene (2 g/L) were added to 1.25 g of lyophilized samples, followed by adding 5 mL of 6 M HCl to bring total volume up to 12.5 mL. The final mixture consisted of 1.2 M HCl in 50% aqueous methanol. The extraction mixture was thereafter heated to 90°C in a steam bath and refluxed for 2 h, allowed to cool in a refrigerator, sonicated for 5 min, and diluted to 50 mL with methanol to form the final extract. The acid hydrolysates methanolic extract was ready for high-performance liquid chromatography (HPLC), inhibition of conjugated diene formation in the linoleic acid assay, TEAC assay, and comet assay.

Polyphenol Assay

Polyphenol content was determined according to the method of Taga *et al.* [20]. Briefly, standard gallic acid and an aliquot of methanolic extract were diluted with an ethanol/water (60:40, v/v) solution containing 0.3% HCl. Two mL of 2% Na_2CO_3 was mixed into each sample of 100 μL and allowed to equilibrate for 2 min before adding 50% Folin-Ciocalteau reagent. Absorbance at 750 nm was measured at room temperature. The standard curve of gallic acid was used to calculate polyphenol levels.

Flavonols Analysis by HPLC

One mL of acid hydrolysates methanolic extract was filtered through a 0.45 μm filter prior to 20 μL being injected into the HPLC. Samples were analyzed with a SpectraSYSTEMUV6000LP Photodiode Array Detection System (Thermo Separation Products, San Jose, USA) and an ODS column (250 × 4.6 mm, 5 μm; YMC, Kyoto, Japan). The mobile phase consisted of methanol–water (30:70, v/v) with 1% formic acid and 100% methanol. The gradient was 25 - 74% methanol in 40 min at a flow rate of 0.75 mL/min. Spectra were recorded at 365 nm for flavonols [19].

Inhibition of Conjugated Diene Formation in Linoleic Acid Emulsion Autoxidation System

The inhibition of conjugated diene formation was determined according to Mitsuda et al. [21]. Briefly, an aliquot of 0.1 mL of diluted plant acidic methanolic extract or blank was added to 2 mL of 10 mM linoleic acidemulsion (pH 6.6), mixed well, and incubated at 37°C for 15 h. A sample of 0.2 mL for 0 and 15 h incubation periods were mixed with 7 mL of 80% methanol, followed by measuring the absorbance at 234 nm.

Trolox Equivalent Antioxidant Capacity (TEAC) Analysis

The total antioxidant capacity of hydrophilic and lipophilic antioxidants was determined using the horseradish peroxidase catalyzed oxidation of 2,2-azino-bis-(3-ethylbenzothiazoline-6-sulfonicacid) (ABTS) [22]. The reaction mixture contained 0.5 mL of 1000 μM ABTS (in ddH$_2$O) and 3.5 mL of 100 μM H$_2$O$_2$ (in ddH$_2$O). The reaction was started by adding 0.5 mL of 44 U/mL peroxidase (in 0.1 M PBS). After 1 h, 0.05

mL of plant acidic methanolic extracts were added to the mixture. After 5 min, absorbance was measured at 730 nm. Trolox (TR) was used as a standard, and the total antioxidant capacity of plant extracts were measured as mM TR equivalent.

Isolated Human Peripheral Blood Lymphocytes

Fasting blood samples were obtained from six donors, including four male and two female healthy non-smokers, 24–48 years old. Fresh venous blood (20–30 mL) was collected in lithium heparin tubes (Becton- Dickinson) from volunteers, and lymphocytes were isolated using a separation solution kit supplemented with Ficoll-Paque Plus lymphocyte isolation sterile solution (Pharmacia Biotech, Sweden) [23]. Cells were harvested within 1 day of taking the blood samples and cultured with AIM V serum-free lymphocyte medium (Gibco Invitrogen, USA) in a humidified atmosphere of 5% CO_2 in air at 37°C for 24 h.

Cell Viability Testing

After culturing, lymphocytes were exposed to each of six different plant acidic methanolic and water extracts. Each lymphocyte was treated with three concentrations of plant acidic methanolic and water extracts (25, 50, and 100 µg/mL) for 30 min at 37°C. DNA damage was induced by exposing lymphocytes to H_2O_2 (10 µM) for 5 min on ice to minimize the possibility of cellular DNA repair after H_2O_2 injury. Cells were centrifuged (100 g for 10 min), washed, and re-suspended in the same medium as the comet assay. All experiments were carried out in triplicate. Cell viability was tested using the tetrazolium/formazan (MTT) assay [24] both prior to and after treatment with plant extracts or H_2O_2.

DNA Single Strand Break Damage Estimation Using the Comet Assay

The standard comet assay was performed as described in Szeto et al. [3], with acidic methanolic and water extracts from these six herbal plants being used for this study. Cultured lymphocytes (10^5 cells/

mL) were embedded in 75 μL of 1% low-melting-point agarose on a microscope slide (precoated with agarose) at 37°C. The gel was allowed to set at 4°C, and cells were lysed for a period of at least 2 h in lysis buffer at 4°C. Cells were then alkaline-unwound, following which electrophoresis was carried out using the electrophoresis buffer at 4°C for 15 min at 25 V with the current adjusted to 300 mA. All steps were conducted under dim light to prevent the occurrence of additional DNA damage. Following electrophoresis, slides were neutralized with neutralization buffer and stained with ethidium bromide. The comet-like images resulting from the extension of DNA were scored as a reflection of the single strand breaks under a fluorescence microscope (Zeiss-Axiovert 100, Zeiss, Germany). Triplicate slides were prepared for each experimental point sample, and 50 comet-like images selected at random per slide were evaluated to determine average DNA damage values. A computerized image analysis system (VisCOMET 1.6, Impuls GmbH, Germany) was employed to determine various comet parameters, and used to analyze DNA damage by tail DNA% [(total brightness of tail area / total brightness of total area) × 100%] and tail moment (tail length × tail DNA%). Inhibition percentage of tail DNA% and tail moment were calculated relative to the 10 μM H_2O_2 treated group.

Statistical Analysis

Data were analyzed by one-way analysis of variance (ANOVA), and the significance between means by the least significant difference (LSD) test. Pearson's linear correlation was also determined. Means of three replicates were reported.

RESULTS

Antioxidant Composition and Antioxidant Activity

Table 1 documents the content of polyphenol in the leaves of tested plants. Polyphenols were significantly abundant in both BA (32.90

mg gallic acid/g DW) and CA (32.03 mg gallic acid/g) compared to other plants. Table 2 presents varied amounts of flavonols ranging from 53.33 to 3200 µg/g DW in the acidic methanolic extract of the studied plants. BA and CA were also rich in myricetin, at levels of 1133.33 and 960.00 µg/g DW, respectively. Morin was present only in MA, CA, and BA plants at a level of 2000.00, 600.00, and 573.33 µg/g DW. Quercetin was abundant in HC (3200.00 µg/g DW), while CA followed at a level of 533.33 µg/g DW. Kaempferol was abundant in CA at a level of 853.33 µg/g DW, but LC and PA did not contain any kaempferol at all. Thus, these species displayed variations in their polyphenol and flavonol levels.

Table 1: The content of polyphenol in tested herbaceous plants

Sample	Ployphenol (mg gallic acid/g DW)
BA	32.90[a]
LC	25.31[b]
MA	21.24[c]
PA	24.31[b]
HC	19.82[d]
CA	32.03[a]

Means with different superscripts [a-d] are significantly different, $p < 0.05$.

DW: dry weight.

Lin *et al.*

Lin *et al.* BMC Research Notes 2013 6:490, doi:10.1186/1756-0500-6-490

Table 2: The content of flavonols in acidic methanolic extracts of tested herbaceous plants

Flavonols (µg/g DW)				
Sample	Myricetin	Morin	Quercetin	Kaempferol
BA	1133.33[a]	573.33[c]	93.33[d]	66.67[c]
LC	320.00[d]	N.D.	N.D.	N.D.
MA	480.00[c]	2000.00[a]	53.33[e]	333.33[b]

PA	253.33[e]	N.D.	213.33[c]	N.D.
HC	146.67[f]	N.D.	3200.00[a]	53.33[c]
CA	960.00[b]	600.00[b]	533.33[b]	853.33[a]

Means within a column with different superscripts ([a~f]) are significantly different, $p < 0.05$. N.D.: not detectable.

Lin et al.

Lin et al. BMC Research Notes 2013 **6**:490, doi:10.1186/1756-0500-6-490

The inhibition of linoleic acid peroxidation was observed to be significantly higher in PA and BA at both 25 and 50 µg/mL of plant extracts (Table 3). Furthermore, significantly higher percentages of conjugated diene inhibition were detected in PA (79.31) and CA (77.61) compared to MA (70.31) at 100 µg/mL of the extract. Hence, each species showed significant differences in inhibition percentages of conjugated dienes at various extract concentrations.

Table 3: Inhibition percentage of conjugated diene formation in the linoleic acid emulsion autoxidation system treated with various concentrations of methanolic acid hydrolysates of herbaceous plants

Inhibition percentage			
Sample	25 µg/mL	50 µg/mL	100 µg/mL
BA	49.14[ab]	55.71[b]	73.72[b]
LC	40.14[c]	52.31[c]	74.69[b]
MA	46.23[b]	56.20[b]	70.31[c]
PA	54.50[a]	68.37[a]	79.31[a]
HC	45.49[b]	59.61[b]	72.99[b]
CA	41.11[c]	58.63[b]	77.61[a]

Means within a column with different superscripts ([a-c]) are significantly different, $p < 0.05$.

Lin *et al.*

Lin *et al.* BMC Research Notes 2013 **6**:490, doi: 10.1186/1756-0500-6-490

Plant extracts from the six species showed antioxidant activities, proving their capacity to scavenge the ABTS radical-cation. The antioxidant activity in methanolic acid hydrolysate extracts of leaf tissues of studied species were expressed in Trolox Equivalent Antioxidant Capacity (TEAC) (Table 4). HC showed a significantly higher TEAC value (231.16 mM) than other species.

Table 4: The TEAC values of acidic methanolic extracts in the investigated herbaceous plants

Samples (100 µg/mL)	TEAC (mM Trolox)
BA	184.61[b]
LC	117.44[e]
MA	146.44[c]
PA	142.49[c]
HC	231.16[a]
CA	132.80[d]

Means within a column with different superscripts ([a-e]) are significantly different, $p < 0.05$.

Lin *et al.*

Lin *et al.* BMC Research Notes 2013 6:490, doi:10.1186/1756-0500-6-490

Effects of Acidic Methanolic and Water Extracts from Herbaceous Plants on H_2O_2-Induced DNA Damage to Lymphocytes

Lymphocytes were exposed to each of three different herbal extracts at three concentrations (25, 50, and 100 µg/mL) for 30 min at 37°C. DNA damage was induced by exposing lymphocytes to H_2O_2 (10 µM)

for 5 min on ice. At two lower levels, no extracts were cytotoxic at the concentrations used, with > 98% of cells remaining viable [25]. Therefore, concentrations only at 25 and 50 µg/mL were chosen for the comet assay. The comet assay was performed to determine the DNA damaging activity of the plants as it is a sensitive method for monitoring single strand DNA breaks at the single cell level. Any DNA damage is represented as tail DNA% and tail moment. The effects of pretreatment of the six tested extracts on 10 µM H_2O_2-induced DNA oxidative damage in human lymphocytes are presented in Figure 1. Tail DNA% demonstrated that MA had a significantly greater level of protection against H_2O_2 exposure than lymphocytes that were exposed to other tested compounds at two doses (25 and 50 µg/mL) (Figure 1A). The maximum protective effect of lymphocyte pretreatment was observed with pretreatment by 25 µg/mL MA, exhibiting 12.43% of tail DNA% compared to the rest of treated samples. Furthermore, at lower concentrations, all tested samples had lower tail DNA%, indicating better inhibition efficacies. The MA extract at the 50 µg/mL was significantly lower than the rest of treated samples, except for HC extract. Tested plants showed at least 707.53 and 1040.63 of tail moment in HC extract at 25 and 50 µg/mL levels compared to the rest of the acidic methanolic extract samples (Figure 1B).

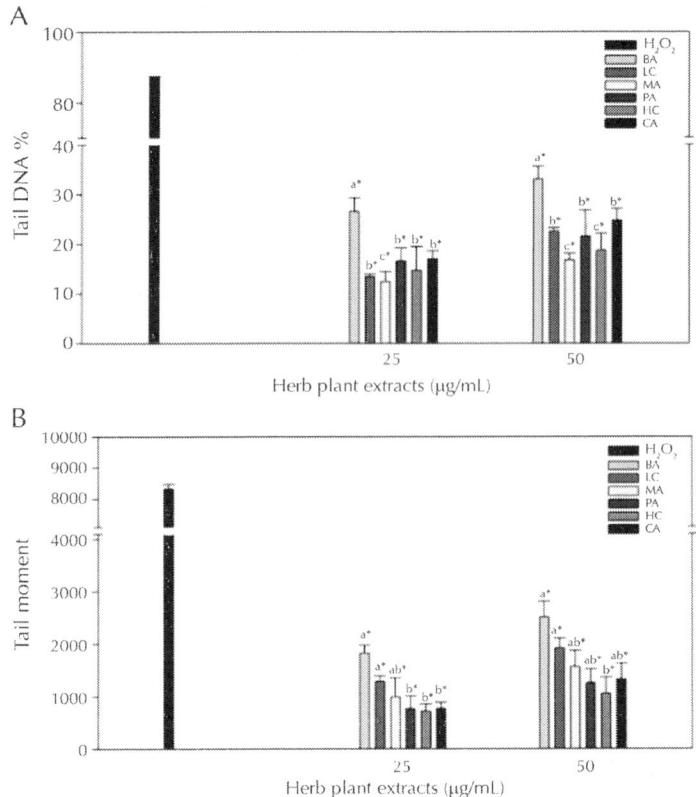

Figure 1: Effects of various acidic methanolic extracts from six herbaceous plants on H2O2-induced DNA damage to lymphocytes. Tail DNA% (A) and tail moment (B) were measured after exposure to tested compounds at 25 and 50 µg/mL of extract. ☐, BA;■, LC; ☐, MA; ■, PA; ■, HC; ■, CA. Values with different letters differ significantly with regard to oxidative damage when comparing different plant extracts; *$p < 0.05$ refers to differences in oxidative damage as compared with 10 µM H_2O_2-alone (■) treatment.

HC had the lowest % tail DNA at 11.14% in 25 µg/mL of water extract (Figure 2A). Both HC (18.36%) and MA (18.25%) extracts at 50 µg/mL had lowest % tail DNA compared to the rest of the water extract of samples. Moreover, HC also had a significantly lower tail moment (1255.40 ~ 1826.10) than the rest of the water extracts at the same doses (Figure 2B). Hence, the DNA damage induced by H_2O_2 was significantly high as compared to the treated extracts, which had 87.26 in tail DNA% and 8328.84 in tail moment.

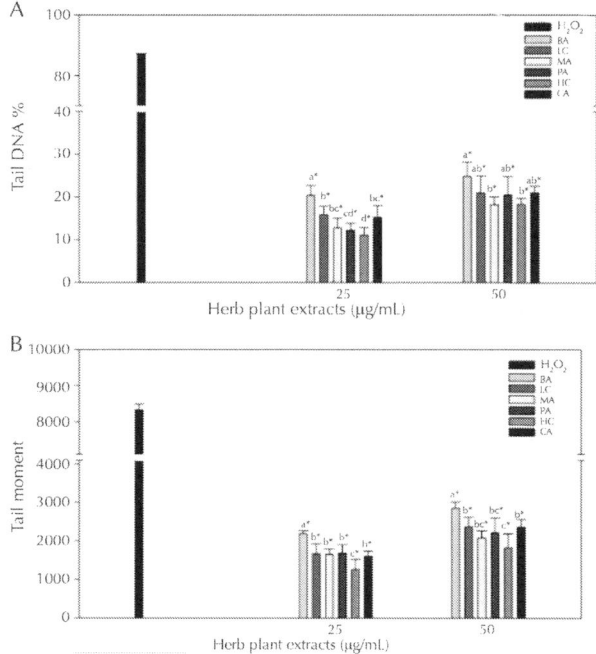

Figure 2: Effects of various water extracts from six herbaceous plants on H2O2-induced DNA damage to lymphocytes. Tail DNA%(A) and tail moment (B) were measured after exposure to tested compounds at 25 and 50 µg/mL of extract. ▨, BA; ▧, LC; ☐▨, PA; ▨, HC; ■, CA. Values with different letters differ significantly with regard to oxidative damage when comparing different plant extracts; $*p < 0.05$ refers to differences in oxidative damage as compared with 10 µM H_2O_2-alone (■) treatment.

DISCUSSION

Antioxidant Composition and Antioxidant Activity

Plant leaves are rich in flavonols and other pigments. BA and CA plants contain higher polyphenol levels than the other plants tested (Table 1). Antioxidant activities are known to increase proportionally to the polyphenol content, mainly due to their redox properties [1].

Among the diverse roles of polyphenols, they protect cell constituents against destructive oxidative damage, thus limiting the risk of various degenerative diseases associated with oxidative stress and tending to be potent free radical scavengers. Their ability to act as antioxidants depends on their chemical structure and ability to donate/accept electrons, thus delocalizing the unpaired electron within the aromatic structure [26]. Phenolic compounds are known as radical scavengers or radical-chain breakers, and they strongly eliminate oxidative free radicals. Quercetin and morin are the principal flavonol constituents in HC and MA plants, respectively (Table 2). These antioxidant compounds may account for the high antioxidant power of the plants in the present study. Quercetin, kaempferol, morin, and myricetin are the most common flavonols, and are the most widely distributed flavonoids in plant leaves. Quercetin, the most abundant flavonoid in the human diet, is an excellent free radical scavenging antioxidant [27]. Polyphenol and flavonol contents found in the extracted plants (Tables 1, 2) were much lower than those in our previous study where purple-leaved sweet potato appeared to have higher contents [28]. A possible reason is the usage of different extraction methods. In fact, different results were obtained from the water and acidic methanolic extracts, and especially from the water extracts. The antioxidant composition and activities of herbal plants cannot be evaluated by a single method due to the complex nature of plants, in which pigments and phytochemicals have specific functions. Therefore, several methods should be employed to evaluate the total antioxidant effects of any plant. Antioxidant compounds presented in plant extracts are therefore multi-functional and their activities and mechanisms of action would largely depend on the composition and conditions of the test system.

Compared to the inhibition percentage of conjugated diene formation in the linoleic acid emulsion autoxidation system of tested samples, PA exhibited relatively higher effectiveness than the others at all extract concentrations (Table 3). The tested vegetables showed >70% inhibition of linoleic acid peroxidation in 100 µg/mL extracts, and PA in particular exhibited the highest inhibition of linoleic acid peroxidation, up to 79.31%. Therefore, all tested plants were effective inhibitors and exhibited better inhibition efficacy at higher concentrations. Previously, we demonstrated that water and methanolic extracts from PA both had higher antioxidant activity, and that the antioxidant activity of PA was equivalent to 10^{-4} M of Trolox in preventing conjugated diene formation during linolic acid peroxidation at 62.5 μg/mL of methanolic extract [29]. The polyphenol content of methanolic extracts was significantly correlated with the delay of the lag phase of low-density lipoprotein (LDL) treated with methanolic extracts. Moreover, the polyphenol content of the methanolic extract of herbaceous plants was significantly correlated with scavenging DPPH radical activity and ferric reducing power [29].

We measured the direct antioxidant activity of acidic methanolic extracts by TEAC assay, reflecting the major mechanisms of antioxidant action for evaluating their relevance to cell protection (Table 3). Jastrzebski et al.[30] reported that prolipid, a mixture of herbs used as a plasma lipid lowering medicine, had strong antioxidant activity. The correlation coefficients between the polyphenols, flavonoids, and TEAC of prolipid water extracts were 0.97 and 0.90, respectively. They concluded that the content of polyphenol in prolipid was the main contributors to the overall antioxidant activity of prolipids. The antioxidant activity of leaf extracts from CA was found to have a direct linear relationship between total phenolic content and total antioxidant activity, indicating that phenolic compounds might be the major contributors to the antioxidant activities of CA extracts [31]. Chung et al.[29] reported that PA, BA, CA, Curled Spearmint, MA, and Mesona had higher total phenolic contents compared to LC and Taiwan lily, and that CA and PA had higher antioxidant activity. In this study, we found that HC and CA contained abundant quercetin while MA and CA were rich in morin and kaempferol, respectively. Additionally, BA and CA had significantly higher levels of myricetin than other tested samples (Table 2). These different pigments may exhibit effective antioxidant activity alone or synergistically, and are a likely cause of cultivar differences. Wang et al. [32] demonstrated that the H donation potential was quercetin > myricetin > morin > kaempferol, indicating that the presence of a 3',4'-catechol moiety in the B ring correlated with high activity. Moreover, the structural peculiarity of di-OH in the B ring obviously rendered quercetin and morin more potent as ROS inhibitors than myricetin and kaempferol, which have tri- and mono-OH in the B ring, respectively. The unclear relationship between antioxidant activity and flavonol extracts indicates that the structure prerequisite to reinforce free radical scavenging activity may vary with the type of free radical. The synergisms among antioxidants make antioxidant activity dependent not only on the concentration, but also might be due to their structures and interactions among antioxidants [33]. The accumulation of flavonoid metabolites in the appropriate target site is probably required to exert their antioxidant activity. The polyphenol-rich plant extracts exhibited distinct cell-free antioxidant activity (TEAC) according to their levels of polyphenol and flavonols, with distinct antioxidant activity strongly accounting for the antioxidant activity of the extracts. HC plants containing 3200 µg/g DW quercetin (Table 2) exhibited the highest TEAC value (231.16 mM) within the tested extracts (Table 4).

Estimation of DNA Single Strand Break Damage from Exposure to Acidic Methanolic and Water Extracts

Quercetin was found to protect against H_2O_2-induced DNA damage in human lymphocytes at 10 µM [34] and at 3.1 to 25 µM [35]. However, it was found to induce DNA damage in human lymphocytes at higher concentrations, such as 100 µM or above [34]. Similarly, myricetin was also found to decrease oxidant-induced DNA damage at 100 µM, although α–tocopherol and β–carotene did not behave similarly. This might be due to the dihydroxy structure of quercetin and myricetin being essential for protecting DNA against hydrogen peroxide [34]. No such hydroxyl groups are present in the tocopherol molecule. This may reflect structure/activity relationships or the localization of the antioxidant relative to free radical generation within cells. Noroozi et al. [36] demonstrated that, in addition to quercetin, kaempferol could also inhibit H_2O_2-induced DNA strand breaks in human lymphocytes. Zhu and Loft [37] reported that aqueous extracts of cooked and autolysed Brussels sprouts decreased DNA strand breaks in human lymphocytes, with the maximum inhibition being 38 and 39% at cooked and autolysed extract levels of 10 µg/mL and 5 µg/mL, respectively, with the inhibition effect decreasing at increasing concentrations up to 100 µg/mL. Quercetin-rich onions showed increased resistance of lymphocytic DNA to ex vivo-induced oxidation [15]. In addition, several types of natural antioxidants, including flavonols and polyphenolic compounds, inhibit adhesion molecule expression and the adhesion of monocytes to endothelial cells, and also suppress cell inflammation, transformation, proliferation, survival, invasion, and angiogenesis [38-40]. Free radicals induce cellular damage and are involved in several human diseases such as cancer, atherosclerosis, and inflammatory disorders, and polyphenols tend to reduce mutagenic activity and oxygen-free radicals [41]. Since the initiation and progress of carcinogenesis involves mutations of DNA, the chemical alteration of DNA bases is believed to be a crucial factor. As a consequence of increased oxidative stress, DNA oxidation damage can occur with ROS, leading to mispairing of DNA bases or DNA strand breaks. ROS are generated endogenously from cellular metabolism and inflammatory responses or by exposure to exogenous agents such as ionizing radiation and xenobiotics [42].

In our study, the inhibition percentages of tested plants ranged from 74.51% (BA) to 91.45% (MA) with acidic methanolic extract concentrations at 25 µg/mL (Figure 1A). MA plants had a value of 985.73 (91.95% inhibition percentage) for tail moment at 25 µg/mL of acidic methanolic extracts (Figure 1B). The results in inhibition percentage of tail DNA% were not similar to the results in inhibition percentage of tail moment among treated samples. The MA plant extract was most effective against DNA single strand breaks in tail DNA%, while HC plant extract was most effective against DNA single strand breaks in tail moment (Figure 1A and 1B). In addition, HC plant water extracts exhibited 11.14% tail DNA% (Figure 2A) and 1255.40 (92.19% inhibition percentage) tail moment at the 25 µg/mL dose (Figure 2B). The inhibition percentage of tail DNA% results was similar to the results of the inhibition percentage of tail moment among treated samples. HC plant extracts not only had the highest Trolox equivalent (Table 4), but were also the most effective against DNA single strand breaks induced by H_2O_2 in human lymphocytes (Figure 1), indicating that it contains polyphenol (19.82 mg gallic acid/g DW), myricetin (146.67 µg/g DW), quercetin (3200.00 µg/g DW), and kaempferol (53.33 µg/g DW) (Tables 1 and 2). To some extent, the observed efficacy of the extracts against DNA damage can be attributed to specific flavonol constituents. The high levels of quercetin and morin are believed to account for the high DNA protective potential of HC and MA since quercetin has also been identified as an efficient reducer of DNA damage in Caco-2 cells [43]. Morin from *Psidium guajava* was effective in increasing cell viability, decreasing ROS levels, and preventing DNA fragmentation upon exposure to high glucose levels in primary rat hepatocyte cultures [44]. The antioxidant activity of polyphenolic compounds in different species showed higher polyphenolic content and antioxidant activity in all species, demonstrating that the tested species are a potent source of novel bioactive compounds with a wide range of medicinal properties. In particular, they have significant free radical scavenging activity. Our present study demonstrates that, among the six investigated species, the higher content of polyphenols, flavonols, and antioxidant properties in HC and MA plants may be the reason for their wide medicinal use. Both species can be used as potent medicinal herbs for novel bioactive compounds with high free radical scavenging activity, and extracts of these plants may been attractive alternative for managing oxidative stress-induced liver injury and drug-induced gastric ulcer [45,46]. Recently, Gargouri et al. [47] demonstrated that quercetin could protect against dimethoate-induced oxidative stress by decreasing lipid peroxidation and protein oxidation, and increasing superoxide dismutase and catalase activities in human lymphocytes. The herbaceous plant extracts in our study may increase antioxidant enzyme activities to protect against H_2O_2-induced DNA damage in human lymphocytes.

CONCLUSIONS

Polyphenol-rich extracts from the tested plants effectively diminish DNA oxidation damage. This preventive effectiveness is attributable to the induction of cellular defenses rather than the radical scavenging activity of polyphenol and flavonols, and might well contribute to the reported health benefits of herbals. The contents of these bioactive compounds in MA and HC extracts can explain their antioxidant activity, and there exists a relationship between the content of polyphenol and flavonol to antioxidant activity. This is the first report suggestion that MA and HC plants have abundant antioxidants with strong antioxidant activity, and consequently can protect DNA in lymphocytes from oxidative damage.

AUTHORS' CONTRIBUTIONS

KHL prepared the extracts and carried out all the experimental process. PYC designed the current project, supervised the work and wrote the manuscript. YYY worked closely with KCL and MYH in the laboratory to carry out the experiments. HFL and HSL evaluated the data and edited the manuscript. CMY participated in statistical analysis. All the authors read and approved the final manuscript.

REFERENCES

1.	Rasineni GK, Siddavattam D, Reddy AR: Free radical quenching activity and polyphenols in three species of Coleus. *J Med Plants Res* 2008, 2:285-291.

2.	McClain DE, Kalinich JF, Ramakrishnan N: Trolox Inhibits apoptosis in irradiated MOLT-4 lymphocytes. *FASEB J* 1995, 9:1345 1354.

3.	Szeto YT, Collins AR, Benzie IF: Effects of dietary antioxidants on DNA damage in lysed cells using a modified comet assay procedure. *Mutat Res* 2002, 500:31-38

4.	Lazzé MC, Pizzala R, Savio M, Stivala LA, Prosperi E, Bianchi L: Anthocyanins protect against DNA damage induced by tertbutyl-hydroperoxide in rat smooth muscle and hepatoma cells. *Mutat Res* 2003, 535:103-115.

5.	Cao G, Sofic E, Prior RL: Antioxidant capacity of tea and common vegetables.*J Agric Food Chem* 1996, 44:3426-3431

6.	Chu YH, Chang CL, Hsu HF: Flavonoid content of several vegetables and their antioxidant activity. *J Sci Food Agr* 2000, 80:561-566.

7.	Farombi EO, Hansen M, Ravn-Haren G, Moller P, Dragsted LO: Commonly consumed and naturally occurring dietary substances affect biomarkers of oxidative stress and DNA damage in healthy rats. *Food Chem Toxicol* 2004, 42:1315-1322.

8.	Taguchi K, Hagiwara Y, Kajiyama K, Suzuki Y: Pharmacological studies of Houttuyniae herba: the anti-inflammatory effect quercitrin. *Yakugaku Zasshi* 1993, 113:327-333

9.	Kim HP, Kim SY, Lee EJ, Kim YC: Zeaxanthin dipalmitate from Lycium chinese has hepatoprotective activity. *Res Commun Mol Pathol Pharmacol* 1997, 97:301-314.

10.	Chen YY, Liu JF, Chen CM, Chao PY, Chang TJ: A study of the antioxidative and antimutagenic effects of Houttuynia cordata

Thunb using an oxidized frying oil-fed model. *J Nutr Sci Vitaminol* 2003, 49:327-333.

11. Geissberger P, Sequin U: Constituents of *Bidens pilosa* L.: Do the components found so far explain the use of this plant in traditional medicine? *Acta Trop* 1991, 48:251-261.

12. Dimo T, Nguelefack TB, Kamtchouing P, Dongo E, Rakotonirina A, Rakotonirina SV:Hyperotensive effects of a methanol extract of Bidens pilosa Linn on hypertensive rats. *C R Acad Sci III* 1999, 322:323-329.

13. Chin HW, Lin CC, Tang KS: The hepatoprotective effects of Taiwan folk medicine ham-hong-chho in rats. *Am J Chin Med* 1996, 24:231-40.

14. Chen YY, Chen CM, Chao PY, Chang TJ, Liu JF: Effects of frying oil and Houttuynia cordata thunb on xenobiotic-metabolizing enzyme system of rodents. *World J Gastroenterol* 2005, 11:389-392.

15. Scalbert A, Johanson IT, Saltmarsh M: Polyphenols: antioxidants and beyond. *Am J Clin Nutr* 2005, 81:215-217.

16. Yen GC, Chuang DY: Antioxidant properties of water extracts from Cassia tora L. in relation to the degree of roasting. *J Agric Food Chem* 2000, 48:2760-2765.

17. Lin KH, Chao PY, Yang CM, Cheng WC, Lo HF, Chang TR: The effects of flooding and drought stresses on the antioxidant constituents in sweet potato leaves. *Bot Stud* 2006, 47:417-426.

18. Liu YL, Tang LH, Liang ZQ, You BG, Yang SL: Growth inhibitory and apoptosis inducing by effects of total flavonoids from Lysimachia clethroides Duby in human chronic myeloid leukemia K562 cells. *J Ethnopharmacol* 2010, 131:1-9.

19. Justesen U, Knuthsen P, Leth T: Quantitative analysis of flavonols, flavones, and flavanones in fruits, vegetables and beverages by high-performance liquid chromatography with photo-diode array and mass spectrometric. *J Chromatogr* 1998, 799:101-110.

20. Taga MS, Miller EE, Pratt DE: Chia seeds as asource of natural lipid antioxidants.*J Am Oil Chem Soc* 1984, 61:928-931.

21. Mitsuda H, Yasumodo K, Iwami K: Antioxidative Action of indole compounds during the autoxidation of linoleic acid. *Eiyo to Shokuryo* 1966, 19:210-214.

22. Re R, Pellegrini N, Evans C: Antioxidant activity applying an improved ABTS radical cation decolorization assay. *Free Rad Biol Med* 1999, 26:1231-1237.

23. Cole J, Green MHL, James SE, Henderson L, Cole H: A further assessment of factors influencing measurements of thioguanine-resistant mutant frequency in circulating T-lymphocytes. *Great Brit Mut Res* 1988, 204:493-507.

24. Cory AH, Owen TC, Barltrop JA, Cory JG: Use of an aqueous soluble tetrazolium/formazan assay for cell growth assays in culture. *Cancer Commun* 1991, 3:207-212

25. Yang YY: *The antioxidative capacity in herb plant extracts and their protection role in DNA oxidative damage of lymphocyte.* Taipei, Taiwan: Chinese Culture University; 2004. [*M.S. Thesis*]

26. Ross JA, Kasum CM: Dietary flavonoids: bioavailability, metabolic effects, and safety. *Annu Rev Nutr* 2002, 22:19-34

27. Villano D, Fernandez-Pachon S, Troncoso AM, Garcia-Parrilla MC: Comparison of antioxidant activity of wine phenolic compounds and metabolites in vitro. *Anal Chim Acta* 2005, 538:391-398

28. Tang SC, Lo HF, Lin KH, Cheng TJ, Yang CM, Chao PY: The antioxidant capacity of extracts from Taiwan indigenous purple-leaved vegetables. *J Taiwan Soc Hort Sci* 2013, 59(1):43-57.

29. Chung AL, Lo H-F, Lin KH, Liu KL, Yang CM, Chao PY: Study on the Antioxidant Activity in Herb Plant Extracts. *J Taiwan Soc Hort Sci* 2013, 59(2):139-152.

30. Jastrzebski Z, Tashma Z, Katrich E, Gorinstein S: Biochemical characteristics of the herb mixture Prolipid as a plant food supplement and medicinal remedy. *Plant Foods Hum Nutr* 2007, 62:145-150.

31. Zaniol MK, Hamid A, Yusof S, Muse R: Antioxidative activity and total phenolic compounds of leaf, root and petiole of four accessions of Centell aasiatica (L). *Urban. Food Chem* 2003, 81:575-581.

32. Wang L, Tu YC, Lian TW, Hing JT, Yen JH, Wu MJ: Distinctive antioxidant and anti-inflammatory effects of flavonols. *J Agric Food Chem* 2006, 54:9798-804

33. Vanderjagt TJ, Ghattas R, Vanderjagt DJ, Glew RH: Comparison of the total antioxidant content of 30 widely used medicinal plants of New Mexico. *Life Sci* 2002, 70:1035-1040

34. Duthie SJ, Collins AR, Duthie GG, Dobson VL: Quercetin and myricetin protect against hydrogen peroxide-induced DNA damage (strand breaks and oxidized pyrimidines) in human lymphocytes. *Mut Res* 1997, 393:223-231.

35. Liu GA, Zheng RL: Protection against damaged DNA in the single cell by polyphenols. *Pharmazie* 2002, 57:852-854

36. Noroozi M, Angerson WJ, Lean ME: Effects of flavonoids and vitamin c on oxidative DNA damage to human lymphocytes. *Am Soc Clin Nutr* 1998, 67:1210-1218.

37. Zhu CY, Loft S: Effects of Brussels sprouts extracts on hydrogen peroxide-induced DNA strand breaks in human lymphocytes. *Food Chem Toxicol* 2001, 39:1191-1197.

38. Moon MK, Lee YJ, Kim JS, Kang DG, Lee HS: Effect of cafeic acid on tumor necrosis factor-alpha-induced vascular inflammation in human umbilical vein endothelial cells. *Biol Pharm Bull* 2009, 32:1371-1377.

39. Li F, Li C, Zhang H, Lu Z, Li Z, You Q, Lu N, Guo Q: A novel flavonoid derivative, inhibits migration and invasion of human breast cancer cells. *Toxico Appl Pharma* 2012, 261:217-226.

40. Chao PY, Huang YP, Hsieh WB: Inhibitive effect of purple sweet potato leaf extract and its components on cell adhesion and inflammatory response in human aortic endothelial cells. *Cell Adh Migr* 2013, 7:237-245.

41. Aviram M: Review of human studies on oxidative damage and antioxidant protection related to cardiovascular diseases. *Free Rad Res* 2000, 33:S85-S87.

42. Bellion P, Digles J, Will F, Janzowski C: Polyphenolic apple extracts: effects of raw material and production method on antioxidant effectiveness and reduction of DNA damage in Caco-2 cell. *J Agr Food Chem* 2010, 58:6636-6642.

43. Schaefer S, Baum M, Eisenbrand G, Dietrich H, Will F, Janzowski C: Polyphenolic apple juice extracts and their major constituents reduce oxidative damage in human colon cell lines. *Mol Nutr Food Res* 2006, 50:24-33.

44. Kapoor R, Kakkar P: Protective role of morin, a flavonoid, against high glucose induced oxidative stress mediated apoptosis in primary rat hepatocytes. *Plos One* 2012, 7(8):e41663. doi:10.1371/journal.pone.0041663

45. Tian L, Shi X, Zhu J, Ma R, Yang X: Chemical composition and hepatoprotective effects of polyphenol-rich extract from Houttuynia cordata tea. *J Agric Food Chem* 2012, 60:4641-4648

46. Londonkar RL, Poddar PV: Studies on activity of various extracts of Mentha arvensis Linn against drug induced gastric ulcer in mammals. *World J Gastrointest Oncol* 2009, 15:82-88.

47. Gargouri B, Mansour RB, Abdallah FB, Elfekih A, Lassoued S, Khaled H: Protective effect of quercetin against oxidative stress caused by dimethoate in human peripheral blood lymphocytes. *Lipids Health Dis* 2011, 10:149-152

Table 3: Composition of selected fatty acids in purslane (Portulaca oleracea) (% of total FA)[a]

Omara-Alwala et al., 1991 [22]				Simopoulos and Salem, 1986 [10]
Fatty acid	Leaf	Stem	Whole plant	Whole plant
18.3-omega-3	41.4–66.4	2.4–5.9	28.4–42.5	47.6
20.5-omega-3	0.8–12.6	18.6–35.5	6.4–21.5	0.1
22.3-omega-3	1.4–3.3	trace	1.0–3.0	—
22.6-omega-3	0.3–6.4	trace	0.6–5.6	—

Results from Omara-Alwala et al., 1991 [22], and Simopoulos and Salem, 1986 [10], expressed as mg of FA per kg or g of net weight.

Leaves of purslane grown both in the controlled growth chamber and in the wild contained higher amount of alpha-linolenic fatty acid (18: 3 w3) than that of spinach leaves. The highest amount (3.41 mg/g) of alpha-linolenic acid was recorded in growth chamber grown purslane, which was seven times higher than that of spinach leaves (0.48 mg/g) (Table 4).

Table 4: Fatty acid profiles in total lipid extracts from leaves of purslane and spinach

Fatty acid	Chamber grown purslane		Wild purslane		Spinach	
	Dry wt%	mg/g fresh wt	Dry wt%	mg/g fresh wt	Dry wt%	mg/g fresh wt
18.0	1.12	0.064	0.95	0.048	0.78	0.007
18.1	4.99	0.016	2.13	0.10	2.04	0.018
18.2	16.99	0.968	13.45	0.70	11.70	0.10
18.3	59.87	3.41	63.78	3.22	53.85	0.48

Source: Simopoulos et al., 1992 [21]

Lipid Content and Fatty Acid Composition

All fractions contained very low lipid content with 0.47% in stems, 0.51% in leaves, and 0.54% in the flowers (Table 5). In general, polyunsaturated fatty acids (PUFAs) were found to be most abundant in all fractions, followed by saturated (SFAs) and monounsaturated fatty acids (MUFAs). The most predominant fatty acids were 18:3n-3 (50%) in the leaf, 18:3n-6 (46%) in the stem, and 18:2n-6 (30% of total fatty acid) in the flowers. ALA content ranged from 149 to 523 mg (100 g sample) in stems and leaves, respectively. An interesting finding in this study was that 18:3n-6 was found at high levels in all fractions, accounting for 46% in stems, 13% in leaves, and 10% in flowers [23].

Table 5: Fatty acid composition of purslane fractions

Fatty acid	Composition (% of total fatty acids)		
	Leaf	Stem	Flower
15:0	0.39	—	1.01
16:0	13.09	16.90	19.30
18:0	2.29	7.75	4.51
Total SFA	16.42	24.64	24.52
16:1	0.54	0.51	0.90
18:1	4.29	3.38	12.30
Total MUFA	4.83	3.89	13.20
18:2n-6	14.46	9.70	30.11
18:3n-6	13.25	45.57	9.68
18:3n-3	49.70	15.62	21.01
20:0	0.21	0.11	0.29
22:0	0.19	0.16	0.10
24:0	1.12	0.31	1.09
Total PUFA	78.75	71.47	62.28
Lipid content (%)	0.51	0.47	0.54

Source: Siriamornpun and Suttajit, 2010 [23].

Antioxidants

The TPC in cultivars of P. oleracea ranged from 127 ± 13 to 478 ± 45 mg GAE/100 g fresh weight of plant. The IC50 ranged from 0.89 ± 0.07 to 3.41 ± 0.41 mg/mL, the AEAC values ranged from 110 ± 14 to 430 ± 32 mg AA/100 g, and the FRAP values ranged from 0.93 ± 0.22 to 5.10 ± 0.56 mg GAE/g [24] (Lim and Quah 2007). DPPH scavenging (IC50) capacity ranged from 1.30 ± 0.04 to 1.71 ± 0.04 mg/mL, while the ascorbic acid equivalent antioxidant activity (AEAC) values were from 229.5 ± 7.9 to 319.3 ± 8.7 mg AA/100 g, the total phenol content (TPC) varied from 174.5 ± 8.5 to 348.5 ± 7.9 mg GAE/100 g, AAC varied from 60.5 ± 2.1 to 86.5 ± 3.9 mg/100 g, and FRAP ranged from 1.8 ± 0.1 to 4.3 ± 0.1 mg GAE/g [16].

Higher amounts of alpha-tocopherol, ascorbic acid, and beta-carotene were observed in the leaves of purslane grown both in the growth chamber and in the wild, compared to the composition of spinach leaves (Table 6). Growth chamber grown purslane contained the highest amount (22.2 mg and 130 mg per 100 g of fresh and dry weight, resp.) of alpha-tocopherol and ascorbic acid (26.6 mg and 506 mg per 100 g of fresh and dry weight, resp.), whereas beta-carotene was slightly higher in spinach.

Table 6: Antioxidant content of purslane and spinach leaves

Name	Alpha-tocopherol		Ascorbic acid		Beta-carotene	
	Mg/100 g (fresh wt.)'	Mg/100 g (dry wt.)	Mg/100 g (fresh wt.)'	Mg/100 g (dry wt.)	Mg/100 (g fresh wt.)'	Mg/100 g (dry wt.)
Chamber grown purslane	22.2	230	26.6	506	1.9	38.2
Wild purslane	8.2	170	23.0	451	2.2	43.5
Spinach	1.8	36	21.7	430	3.3	63.5

Source: Simopoulos et al, 1992 [21].

Vitamin C (ascorbic acid) and beta-carotene have been reported to possess antioxidant activity, because of their ability to neutralize free radicals, and have the potential to prevent cardiovascular disease and cancer [25]. Leaves had the highest content of beta-carotene, ascorbic acid, and DPPH, followed by flowers and stems (Table7). Thai wild purslane contained almost 10 time's higher beta-carotene and ascorbic acid [3, 4, and 26] content than other varieties. The beta-carotene content in the leaf was two times higher than in the stems and slightly higher than in the flowers. This finding is in agreement with the data on Australian purslane, where the beta-carotene content in the leaf was higher than in the stem [3]. Purslane is amongst the group of plants with high oxalate contents. Melatonin is a ubiquitous and versatile molecule that exhibits most of the desirable characteristics of a good antioxidant [27]. The oxalate content of purslane leaves was reported as 671–869 mg/100 g fresh weight [28, 29].

Table 7: Ascorbic acid and beta-carotene content and 1, 1-diphenyl-2-picryl-hydrazyl (DPPH) radical-scavenging activity of different parts of Thai purslane

Plant parts	Antioxidant compound content		
	Beta-carotene (mg g^{-1} sample)	Ascorbic acid (mg g^{-1} sample)	DPPH (%)
Leaf	0.58	3.99	76.71
Stem	0.29	2.27	90.11
Flower	0.55	2.32	91.01

CONCLUSIONS

As a significant source of omega-3 oils, P. oleracea could yield considerable health benefits to vegetarian and other diets where the consumption of fish oils is excluded. Scientific analysis of its chemical components has shown that this common weed has uncommon nutritional value, making it one of the potentially important foods for the future. Presence of high content of antioxidants (vitamins A and C, alpha-tocopherol, beta-carotene, and glutathione) and omega-3 fatty acids and its wound healing and antimicrobial effects as well as its traditional use in the topical treatment of inflammatory conditions suggest that purslane is a highly likely candidate as a useful cosmetic ingredient.

REFERENCES

1. M. D. Kamal-Uddin, A. S. Juraimi, M. Begum, M. R. Ismail, A. A. Rahim, and R. Othman, "Floristic composition of weed community in turf grass area of west peninsular Malaysia," International Journal of Agriculture and Biology, vol. 11, no. 1, pp. 13–20, 2009.

2. M. K. Uddin, A. S. Juraimi, M. R. Ismail, and J. T. Brosnan, "Characterizing weed populations in different turfgrass sites throughout the Klang Valley of western Peninsular Malaysia," Weed Technology, vol. 24, no. 2, pp. 173–181, 2010.

3. L. Liu, P. Howe, Y.-F. Zhou, Z.-Q. Xu, C. Hocart, and R. Zhang, "Fatty acids and β-carotene in Australian purslane (Portulaca oleracea) varieties," Journal of Chromatography A, vol. 893, no. 1, pp. 207–213, 2000.

4. A. P. Simopoulos, H. A. Norman, and J. E. Gillaspy, "Purslane in human nutrition and its potential for world agriculture," World Review of Nutrition and Dietetics, vol. 77, pp. 47–74, 1995.

5. I. Oliveira, P. Valentão, R. Lopes, P. B. Andrade, A. Bento, and J. A. Pereira, "Phytochemical characterization and radical scavenging activity of Portulaca oleraceae L. leaves and stems," Microchemical Journal, vol. 92, no. 2, pp. 129–134, 2009.

6. A. P. Simopoulos, "Omega-3 fatty acids and antioxidants in edible wild plants," Biological Research, vol. 37, no. 2, pp. 263–277, 2004.

7. I. Gill and R. Valivety, "Polyunsaturated fatty acids. Part 1: occurrence, biological activities and applications," Trends in Biotechnology, vol. 15, no. 10, pp. 401–409, 1997.

8. J. Whelan and C. Rust, "Innovative dietary sources of n-3 fatty acids," Annual Review of Nutrition, vol. 26, pp. 75–103, 2006.

9. P. J. Nestel, "Polyunsaturated fatty acids (n-3, n-6)," The American Journal of Clinical Nutrition, vol. 45, no. 5, pp. 1161–1167, 1987.

10. A. P. Simopoulos and N. Salem Jr., "Purslane: a terrestrial source of omega-3 fatty acids," The New England Journal of Medicine, vol. 315, no. 13, p. 833, 1986.

11. U. R. Palaniswamy, R. J. McAvoy, and B. B. Bible, "Stage of harvest and polyunsaturated essential fatty acid concentrations in purslane (Portulaca oleraceae) leaves," Journal of Agricultural and Food Chemistry, vol. 49, no. 7, pp. 3490–3493, 2001.

12. I. Yazici, I. Türkan, A. H. Sekmen, and T. Demiral, "Salinity tolerance of purslane (Portulaca oleraceae L.) is achieved by enhanced antioxidative system, lower level of lipid peroxidation and proline accumulation," Environmental and Experimental Botany, vol. 61, no. 1, pp. 49–57, 2007.

13. M. K. Uddin, A. S. Juraimi, M. A. Hossain, F. Anwar, and M. A. Alam, "Effect of salt stress of Portulaca oleracea on antioxidant properties and mineral compositions," Australian Journal Crop Science, vol. 6, pp. 1732–1736, 2012.

14. A. N. Rashed, F. U. Afifi, and A. M. Disi, "Simple evaluation of the wound healing activity of a crude extract of Portulaca oleracea L. (growing in Jordan) in Mus musculus JVI-1," Journal of Ethnopharmacology, vol. 88, no. 2-3, pp. 131–136, 2003.

15. A. Danin and J. A. Reyes-Betancort, "The status of Portulaca oleracae in the Canary Islands," Lagascalia, vol. 26, pp. 71–81, 2006.

16. M. K. Uddin, A. S. Juraimi, M. E. Ali, and M. R. Ismail, "Evaluation of antioxidant properties and mineral composition of Portulaca oleracea (L.) at different growth stages," International Journal Molecule Science, vol. 13, pp. 10257–10267, 2012.

17. A. P. Simopoulos, "Evolutionary aspects of diet, essential fatty acids and cardiovascular disease,"European Heart Journal, vol. 3, pp. D8–D21.

18. S. Siriamornpun and M. Suttajit, "Microchemical components and antioxidant activity of different morphological parts of thai wild purslane (Portulaca oleracea)," Weed Science, vol. 58, no. 3, pp. 182–188, 2010.

19. M. A. Dkhil, A. E. A. Moniem, S. Al-Quraishy, and R. A. Saleh, "Antioxidant effect of purslane (Portulaca oleracea) and its mechanism of action," Journal of Medicinal Plant Research, vol. 5, no. 9, pp. 1589–1593, 2011.

20. A. E. Abdel-Moneim, M. A. Dkhil, and S. Al-Quraishy, "The redox status in rats treated with flaxseed oil and lead-induced hepatotoxicity," Biological Trace Element Research, vol. 143, no. 1, pp. 457–467, 2011.

21. A. P. Simopoulos, H. A. Norman, J. E. Gillaspy, and J. A. Duke, "Common purslane: a source of omega-3 fatty acids and antioxidants," Journal of the American College of Nutrition, vol. 11, no. 4, pp. 374–382, 1992.

22. T. R. Omara-Alwala, T. Mebrahtu, D. E. Prior, and M. O. Ezekwe, "Omega-three fatty acids in purslane (Portulaca oleracea) Tissues," Journal of the American Oil Chemists› Society, vol. 68, no. 3, pp. 198–199, 1991.

23. S. Siriamornpun and M. Suttajit, "Microchemical components and antioxidant activity of different morphological parts of thai wild purslane (Portulaca oleracea)," Weed Science, vol. 58, no. 3, pp. 182–188, 2010.

24. Y. Y. Lim and E. P. L. Quah, "Antioxidant properties of different cultivars of Portulaca oleracea," Food Chemistry, vol. 103, no. 3, pp. 734–740, 2007.

25. V. A. Rifici and A. K. Khachadurian, "Dietary supplementation with vitamins C and E inhibits in vitro oxidation of lipoproteins," Journal of the American College of Nutrition, vol. 12, no. 6, pp. 631–637, 1993.

26. M. A. Alam, A. S. Juraimi, M. Y. Rafii et al., "Evaluation of antioxidant compounds, antioxidant activities and mineral composition of 13 collected purslane (Portulaca oleracea L.) accessions," BioMed Research International, vol. 2014, Article ID 296063, 10 pages, 2014.

27. A. Galano, D. X. Tan, and R. J. Reiter, "Melatonin as a natural ally against oxidative stress: a physicochemical examination," Journal of Pineal Research, vol. 51, no. 1, pp. 1–16, 2011.

28. A. P. Simopoulos, D.-X. Tan, L. C. Manchester, and R. J. Reiter, "Purslane: a plant source of omega-3 fatty acids and melatonin," Journal of Pineal Research, vol. 39, no. 3, pp. 331–332, 2005.

29. A. I. Mohamed and A. S. Hussein, "Chemical composition of purslane (Portulaca oleracea)," Plant Foods for Human Nutrition, vol. 45, no. 1, pp. 1–9, 1994.

A Study of Antioxidant Activity, Enzymatic Inhibition and in Vitro Toxicity of Selected Traditional Sudanese Plants with Anti-Diabetic Potential

Yasmin Hilmi[1,] Muna F Abushama[1,] Haidar Abdalgadir[2,] Asaad Khalid[2,] and Hassan Khalid[1]

[1]Khartoum College of Medical Sciences, Khartoum, Sudan
[2]Medicinal and Aromatic Plants Research Institute, National Centre for Research, Khartoum, Sudan

ABSTRACT

Background

Diabetes mellitus is a chronic metabolic disease with life-threatening complications. Despite the enormous progress in conventional medicine and pharmaceutical industry, herbal-based medicines are still a common practice for the treatment of diabetes. This study evaluated ethanolic and aqueous extracts of selected Sudanese plants that are traditionally used to treat diabetes.

Methods

Extraction was carried out according to method described by Sukhdev et. al. and the extracts were tested for their glycogen phosphorylase inhibition, Brine shrimp lethality and antioxidant activity using (DPPH) radical scavenging activity and iron chelating activity. Extracts prepared from the leaves of *Ambrosia maritima,* fruits of *Foeniculum vulgare* and *Ammi visnaga,* exudates *of Acacia Senegal,* and seeds of *Sesamum indicum* and *Nigella sativa.*

Results

Nigella sativa ethanolic extract showed no toxicity on Brine shrimp Lethality Test, while its aqueous extract was toxic. All other extracts were highly toxic and ethanolic extracts of *Foeniculum vulgare* exhibited the highest toxicity. All plant extracts with exception of *Acacia senegal* revealed significant antioxidant activity in DPPH free radical scavenging assay.

Conclusions

These results highly agree with the ethnobotanical uses of these plants as antidiabetic. This study endorses further studies on plants investigated, to determine their potential for type 2 diabetes management. Moreover isolation and identification of active compounds are highly recommended.

BACKGROUND

Diabetes mellitus (DM) is a disease with severe complications and major health/economic impacts. It is a leading cause of morbidity and mortality worldwide, with an estimated 346 million adults being affected in year 2011. WHO projects that diabetes death will increase by two- thirds between 2008 and 2030 [1]. The prevalence of diabetes for all age-groups worldwide was estimated to be 2.8% in 2000 and 4.4% in 2030 [2].

In Sudan diabetes is an increasingly important problem, being responsible for 10% of hospital admissions and mortality [3]. Recently, an increase in incidence of DM has been observed especially among urbanized population indicating that diabetes mellitus is emerging as an important health problem [4]. The results of a small-scale study carried out in 1996 indicated that diabetes population in Sudan is at around one million, 90% of them have type 2 diabetes. It also showed a prevalence of 3.4% of type 2 DM [5].

Nature is an extraordinary source of medicines. The use of traditional medicines and medicinal plants in most developing countries as therapeutic agents for the maintenance of good health has been widely observed. The World Health Organization estimated that 80% of the populations of developing countries rely on traditional medicines, mostly plant drugs, for their primary health care needs. Diabetes is an example of a disease that has been treated with plant medicines. Research conducted in the last few decades on plants used traditionally for treatment of diabetes has shown antidiabetic properties [6].

To date, more than 1200 flowering plants have been claimed to have antidiabetic properties. Among them, only one-third have been scientifically studied and documented in around 460 publications [7].

The Sudanese flora has a vast variety of medicinal plants, which are traditionally used for their antidiabetic property. However, careful assessment including sustainability of such herbs, seasonal variation in activity of phyto-constituents, metal contents of crude herbal anti-diabetic drugs, thorough toxicity study and cost effectiveness is required for their popularity. These efforts may justify the role of novel traditional medicinal plants having anti-diabetic potentials.

Herbal drugs are considered free from side effects than synthetic one. They are less toxic, relatively cheap and popular [8]. However cytotoxicity of these plants needs to be monitored.

Brine shrimp bioassays (BSLT) offer a quick, simple and cost-efficient way of testing the toxicity of plant extracts. It has been developed for screening, fractionation and monitoring of biologically active natural products [9, 10].

Oxidative stress has been implicated in the development of many pathophysiological conditions including diabetes. Oxidative stress takes place due to the disturbance of the balance between the formation of reactive oxygen species (ROS) and the defense provided by cellular antioxidants. Medicinal plants provide a natural source of antioxidants that have been used worldwide for treatment of many diseases [6].

The inhibition of glycogen phosphorylase has been used as one method for treating type 2 diabetes. Since glucose production in the liver has been shown to increase in type 2 diabetes patients, inhibiting the release of glucose from the liver's glycogen's supplies appears to be a valid approach [11, 12].

In the present study we assessed the antioxidant and hypoglycemic activities of ethanolic and aqueous extracts of *Ambrosia maritima, Ammi visnaga, Acacia senegal, Sesamum indicum, Nigella sativa, Foeniculum vulgare* of Sudanese origin. Screening of toxicity of these plants using brine shrimp lethality test, is also investigated.

METHODS

Plant Material

Plants collected from Khartoum local market, identified and authenticated by Dr. Haider Abdelgadir, Herbarium Curator. Herbarium material was deposited at The Medicinal & Aromatic Plants Research Institute (MAPRI), Khartoum, Sudan. Table 1 shows the tested plants, their parts used and medicinal uses.

Table 1: Tested plants

Plant name	Family	Parts used	Reported medicinal uses[17-20]
Acacia senegal	Fabaceae—Mimosoidae	Fruits	Treatment of diabetes and chronic renal failure. Stem exudates (gums) are used as demulcents and against diarrhoea and ulcers
Ambrosia maritima	Asteraceae	Leaves	Hepatoprotective and Molluscicidal, The herbs are used in treatment of urinary tract infections and elimination of kidneystones, whereas the leaves are used as anti-diabetic and anti-hypertensive
Ammi visnaga	Apiaceae	Fruits	Used for renal urethra stones, smooth muscle relaxant
Foeniculum vulgare	Apiaceae	Fruits	Carminative, flatulence and digestive. It is also used in veterinary medicine
Nigella sativa	Ranunculaceae	Seeds	Treatment of diabetes, hypertension, abdominal ulcers, prostate gland and lung injury(Cuneyt Tayman)
Sesamum indicum	Pedaliaceae	Seeds	Treatment, for cough and cold. also used as nutritive, laxative, demulcent and emollient propertie

Hilmi *et al.*

Hilmi *et al.* BMC Complementary and Alternative Medicine 2014 14:149, doi:10.1186/1472-6882-14-149

Preparation of Plant Extracts

Extraction was carried out according to method described by Sukhdev *et. al.*[13].

Preparation of Ethanolic Extract

Specific weight of each plant sample was extracted by soaking in 96% ethanol for about seventy two hours with daily filtration and evaporation. Solvent was evaporated under reduced pressure to dryness using rotary evaporator apparatus and the extract combined together. The yield percentage was calculated as followed:

Weight of extract/weight of sample×100

Preparation of the Aqueous Extract

About 50–100 g of each plant sample was soaked in 500 ml hot distilled water, and left till cooled down with continuous stirring at room temperature. Extract was then filtered and freeze dried. Yield percentage was calculated.

Weight of extract obtained/weight of plantsample×100

Brine Shrimp Lethality Test

Artemia salina (shrimp eggs) was placed in natural sea water, and eggs hatched within 48 hrs, providing a large number of larvae (nauplii). The tested sample (20 mg) was dissolved in 2 ml of ethanol. From this solution 5, 50 and 500 µl were transferred to vials (triplicate for each concentration), forming concentrations of 10, 100 and 1000 µg/ml respectively. The solvent was allowed to evaporate overnight. Volume was made to 5 ml with seawater. 10 larvae were placed in each vial using a Pasteur pipette. Vials were incubated at 25–27°C for 24 hrs under illumination. Etoposide (7.4625 µg/ml) was used as positive control, and number of survived larvae were counted. Data was analyzed by Finney Probit Analysis computer program to determine LC_{50} values with 95% confidence intervals [9].

Antioxidant Activity Assays

DPPH Radical Scavenging Assay

The DPPH radical scavenging was determined according to the modified method of Shimada *et al.*[14] in 96-wells plate, the test samples were allowed to react with 1,1-Diphenyl-2-picryl-hydrazyl stable free radical (DPPH) for half an hour at 37°C. The concentration of DPPH was kept as 300 µM. The test samples were dissolved in DMSO while DPPH was prepared in ethanol. After incubation, decrease in absorbance was measured at 517 nm using multiple reader spectrophotometer. Percentage radical scavenging activity by samples was determined in comparison with a DMSO treated control group. All tests and analysis were run in triplicate. Vitamin C was used as positive control

Iron Chelating Activity Assay

The iron chelating ability was determined according to the modified method of Dinis *et al.*[15]. The Fe^{+2} were monitored by measuring the formation of ferrous ion-ferrozine complex. The experiment was carried out in 96 microtiter plate. The plant extracts were mixed with $FeSO_4$. The reaction was initiated by adding 5 mM ferrozine. The mixture was shaken and left at room temperature for 10 min. The absorbance was measured at 562 nm. EDTA was used as standard, and DMSO as control. All tests and analysis were run in triplicate.

Glycogen Phosphorylase Enzyme Assays

Glycogen phosphorylase a (from rabbit muscle), glycogen, glucose-1-phosphate, malachite green, and ammonium molybdate were purchased from the Sigma–Aldrich Corporation. Reagents and solvents were obtained from commercial suppliers and used without further purification. Solvents used were AR grade. The enzymatic inhibition of phosphorylase activity was monitored using multiskan spectrum (Thermo-Scientific) based on the published methods. In brief, GPa activity was measured in the direction of glycogen synthesis by the release of phosphate from glucose-1-phosphate. Each compound was dissolved in DMSO and diluted at different concentrations for IC50 determination. The enzyme was added into the 100 µL buffer with compounds dissolved in containing 50 mM Hepes (pH 7.2), 100 mM KCl, 2.5 mM MgCl2, 0.5 mM glucose-1-phosphate, and 1 mg/mL glycogen in 96-well microplates (costar). After the addition of 150 µL of 1 M HCl containing 10 mg/mL ammonium molybdate and 0.38 mg/mL malachite green, reactions were run at 25°C for 20 min. And then the phosphate absorbance was measured at 620 nm [16].

RESULTS AND DISCUSSION

Different medicinal systems that have been discovered as natural hypoglycemic medicine came from the virtue of traditional knowledge and have been used in many countries [7, 21, 22].

Many herbal extracts are currently traditionally used in Sudan for the treatment of diabetes. However, such medicinal plants have not gained much importance as medicines due to the lack of sustained scientific evidence. In the present study, aqueous and ethanolic extracts of six indigenous antidiabetic medicinal plants from Sudan were studied. The rationale for performing extractions from polar to non-polar solvents is to confirm and validate the inhibitory activity in the aqueous extractions performed in the traditional manner as well as to search for newer, more potent inhibitory compounds in the organic solvents. Since the Sudanese population has long been using all these plants for food and medicinal purposes, they form a part of the local pharmacopoeia.

Increasing evidence in both experimental and clinical studies suggests that oxidative stress plays a major role in the pathogenesis of both types of DM. Free radicals are formed disproportionately in diabetes by glucose oxidation, nonenzymatic glycation of proteins, and the subsequent oxidative degradation of glycated proteins. Abnormally high levels of free radicals and the simultaneous decline of antioxidant defense mechanisms can lead to damage of cellular organelles and enzymes, increased lipid peroxidation, and development of insulin resistance. These consequences of oxidative stress can promote the development of complications of DM. Changes in oxidative stress biomarkers, and use of antioxidants is an important trend in the treatment of DM [23].

In the present work we studied the anti-oxidant activity of ethanolic and aqueous extracts of investigated plants. Samples will be considered to have high or significant antioxidant capacity with IC50 < 50 µg/ml (extract) or IC50 < 10 µg/ml (compounds), moderate antioxidant capacity with 50 < IC50 < 100 µg/ml (extract) or 10 < IC50 < 20 µg/ml (compounds) and low antioxidant capacity with IC50 > 100 µg/ml (extract) or IC50 > 20 µg/ml (compounds) [24]. *Ambrosia maritima, Ammi visnaga* and *Foeniculum vulgare* exhibited high antioxidant activity in DPPH free radical scavenging assay with IC50 of 36 µg/ ml of ethanol extract of *Ambrosia maritima*, 41 µg/ ml of ethanol extract of *Ammi visnaga,* 47 µg/ml of aqueous extract of *Ammi visnaga,* 49 µg/ml of ethanol extract of *Foeniculum vulgare* and 31 µg/ml of aqueous extract of *Foeniculum vulgare.* This may support the traditional usage of these plants to improve complications such oxidative stress that caused by DM as well as many other diseases. A study in Egypt by Abu Zid and coworkers showed a moderate antioxidant activity of aqueous extract of *M. ambrosia* [25]. Results of *Achyranthes aspera* extracts in alloxan-treated mice revealed significant anti-hyperglycemic activity that may be mediated by diminished oxidative stress [26]. A study of *Eucalyptus globulus, Salvia officinalis* growing in Algeria and *Guiera senegalensis* growing in Sudan demonstrated that the 96% alcoholic leaf extracts had a significant blood-glucose lowering potential in glucose loaded rats with minimum toxicity [27]. Swanston and coworkers reported that agrimony, alfalfa, coriander, eucalyptus and juniper, can retard the development of streptozotocin diabetes in mice [28]. In another study, ethanolic crude extract of *Sorbus decora* demonstrates both anti-hyperglycemic and insulin-sensitizing activity *in vivo*, thereby confirming anti-diabetic potential and validating traditional medicine [29]. *Trigonella foenum-graecum, Atriplex halimus, Olea europaea, Urtica dioica, Allium sativum, Allium cepa,Nigella sativa,*and *Cinnamomum cassia* were tested for their antidiabetic properties. Results indicated that the observed anti-diabetic properties of these plants are mediated, at least partially, through regulating GLUT4 translocation [30].

Glycogen phosphorylase inhibition has been used as one method for treating type 2 diabetes[11,12]. Results of the current study did not show any significant inhibition of glycogen phosphorylase, but extracts of these plants may act on one of other enzymatic reactions that are involved in carbohydrate metabolism and improved glucose homeostasis.

All aqueous extracts showed significantly high toxicity on Brine shrimp Lethality Test, while*Foeniculum vulgare* showed moderate toxicity. Ethanolic extract of *Nigella sativa* showed no toxicity while all other ethanolic extracts exhibited high toxicity. Ethanolic extracts of *Foeniculum vulgare*exhibited the highest toxicity. These statistical consideration are based on the published work by Bussmann and coworkers. They stated that LC_{50} values below 249 µg/ml are considered as highly toxic, 250–499 µg/ml as median toxicity and 500–1000 µg/ml as light toxicity. Values above 1000 µg/ml are regarded as non-toxic [31]. These results could be very useful as preliminary data in the search for new antitumor compounds from the Sudanese market flora. All results for antioxidant activities, glycogen phosphorylase inhibition and cytotoxicity are shown in Table 2.

Table 2: Antioxidant activity, enzymatic inhibition and cytotoxicity of selected Sudanese medicinal plants

Plant	Extract	DPPH radical scavenging assay %	Iron chelating assay %	Inhibition % of glycogen phosphorylase (5mg/ml)	Brine shrimp lethality (LC 50)
Acacia Senegal	Ethanolic	Not Active	Not Active	0	83.8716
	Aqueous	Not Active	Not Active	0	17.9948
Ambrosia maritima	Ethanolic	60.8 ± 0.04	Not Active	2.2	39.7866
	Aqueous	21.2 ± 0.02	Not Active	0	10.6353
Ammi visnaga	Ethanolic	52.4 ± 0.03	Not Active	0	8.1217
	Aqueous	52.4 ± 0.03	2.5 ± 0.03	0	32.6273
Foeniculum vulgare	Ethanolic	60.7 ± 0.06	3.6 ± 0.05	0	0.012
	Aqueous	69.4 ± 0.003	Not Active	0	893.97
Nigella sativa	Ethanolic	47 ± 0.02	6.3 ± 0.02	0	11684.6
	Aqueous	19.3 ± 0.01	43.5 ± 0.04	0	122.268
Sesamum indicum	Ethanolic	Not Active	Not Active	8.2	61.85
	Aqueous	40.3 ± 0.01	23.2 ± 0.02	0	1.7

Hilmi *et al.*

Hilmi *et al. BMC Complementary and Alternative Medicine* 2014 14:149, doi: 10.1186/1472-6882-14-149

CONCLUSIONS

In conclusion these results revealed the significant antioxidant activity of the investigated plants extracts and may explain their role in altering the oxidative stress and management of diabetes mellitus. Furthermore the high toxicity of many extracts tested in this study suggests their antitumor potential and provides an avenue to explore the bioactive components of plant extracts. Studies should be directed towards drug industry by identification of single chemical compounds, and dosage use has to be monitored.

AUTHORS' CONTRIBUTIONS

YH participated in the study design and coordination, carried out the toxicity assay, drafted the manuscript and rewrote the final one. MA participated in the design of the study, toxicity assay and helped to draft the manuscript. HA conceived of the study, and participated in its design and coordination. AK participated in the enzymatic inhibition study and antioxidant activity. HK supervised part of the study and reviewed the manuscript. All authors read and approved the final manuscript.

ACKNOWLEDGEMENTS

The authors are grateful to the Medicinal & Aromatic Plants Research Centre (Khartoum, Sudan) for providing necessary laboratory facilities. We thank Mr. Muddathir Alhassan for his technical contribution in extraction methods. We thank Ms. Fatima Elfatih for her technical assistance in the enzymatic inhibition assays and Mr. Eltayeb Fadul for carrying out the IC50 analysis.

REFERENCES

1.   *WHO Diabetes Fact Sheet N 312 September (2012)*. http://www. who.int/mediacentre/factsheets/fs312/en/index.html

2.   Wild S, Bchir M, Roglic G, Green A, Sicree R, King H: Global Prevalence of Diabetes Estimates for the year 2000 and projections for 2030. *Diabetes Care* 2004, 27:1047-1053.

3.   Ahmed AM, Ahmed NH, Abdalla ME: Pattern of hospital mortality among diabetic patients in Sudan. *Pract Diabetes Int* 2000, 17:41-43.

4.   Ahmed AM: Diabetes mellitus in Sudan: size of the problem and possibilities of efficient care. *Pract Diabetes Int* 2001, 18:324-327.

5.   Elbagir MN, Eltom MA, Elmahadi EM, Kadam IM, Berne C: A population-based study of the prevalence of diabetes and impaired glucose tolerance in adults in northern Sudan. *Diabetes Care* 1996, 19(10):1126-1128.

6.   Abou El-Soud N, El-Laithy N, El-Saeed G, Wahby M, Khalil M, Morsy F, Shaffie N:Antidiabetic Activities of *Foeniculum Vulgare* Mill. Essential Oil in Streptozotocin-Induced Diabetic Rats. *Macedonian J Med Sci* 2011, 4(2):139-146.

7.   Chang CL, Lin Y, Bartolome AB, Chen YC, Chiu SC, Yang WC: Herbal Therapies for Type 2 Diabetes Mellitus: Chemistry, Biology, and Potential Application of Selected Plants and Compounds. *Evidence-Based Complement Altern Med* 2013, 33.

8.   Gupta R, Bajpai KG, Johri S, Saxena AM: An overview of Indian novel traditional medicinal plants with antidiabetic potentials. *Afr J Trad CAM* 2008, 5(1):1-17.

9.   McLaughlin JL, Rogers LL, Anderson JE: The Use of Biological Assays to Evaluate Botanicals. *Drug Inf J* 1998, 32:513-524.

10.   Bastos MLA, Lima MRF, Conserva LM, Andrade VS, Rocha EMM, Lemos RPL: Studies on the Antimicrobial Activity and Brine Shrimp Toxicity of Zeyheria tuberculosa (Vell.) Bur. (Bignoniceace) Extracts and Their Main Constituents. *Ann Clin Microbiol Antimicrob* 2009, 8:16-20.

11. Somsák L, Nagya V, Hadady Z, Docsa T, Gergely P: Glucose analog inhibitors of glycogen phosphorylases as potential antidiabetic agents: recent developments. *Current Pharmacological Design* 2003, 9(15):1177-1189.

12. Moller DE: New drug targets for type 2 diabetes and the metabolic syndrome. *Nature* 2001, 414(6865):821-827.

13. Sukhdev SH, Suman PSK, Gennaro L, Dev DR: Extraction technologies for medicinal and aromatic plants. 2008. [*Chapter 1. United Nations Industrial Development Organization and the International Centre for Science and High Technology, Italy*]

14. Shimada K, Fujikawa K, Yahara K, Nakamura T: Antioxidative properties of xanthan on the antioxidation of soybean oil in cyclodextrin emulsion. *J Agric Food Chem* 1992, 40(6):945-948.

15. Dinis TCP, Madeira VMC, Almeida LM: Action of phenolic derivates (acetoaminophen, salycilate and 5-aminosalycilate) as inhibitor of membrane lipid peroxidation and as peroxyl radical scavengers. *Arch Biochem Biophys* 1994, 315:161-169.

16. Martin WH, Hoover DJ: Discovery of a human liver glycogen phosphorylase inhibitor that lowers blood glucose in vivo. *Proc Natl Acad Sci U S A* 1998, 95(4):1776-1781.

17. Khalid H, Abdella W, Abdelgadir H, Opatz T, Efferth T: Gems from traditional north-African medicine: medicinal and aromatic plants from Sudan Nat. *Prod Bioprospect* 2012, 2:92-103.

18. Vanachayangkul P, Chow N, Khan SR, Butterweck V: Prevention of renal crystal deposition by an extract of *Ammi visnaga* L. and its constituent's khellin and visnagin in hyperoxaluric rats. *Urol Res* 2011, 39(3):189-195.

19. Ghanem MTM, Radwan HMA, Mahdy EM, Elkholy YM, Hassanein HD, Shahat AA: Phenolic compounds from *Foeniculum vulgare* (Subsp. *Piperitum*) (Apiaceae) herb and evaluation of hepatoprotective antioxidant activity. *Pharmacognosy Res* 2012, 4(2):104-108.

20. Tayman C, Cekmez F, Kafa I, Canpolat F, Cetinkaya M, Tonbul A, Uysal S, Tunc T, Sarici S: Protective Effects of Nigella sativa Oil in Hyperoxia-Induced Lung Injury. *Arch Bronconeumol* 2013, 49(1):15-21.

21. Baldé NM, Youla A, Baldé MD, Kaké A, Diallo MM, Baldé MA, Maugendre D: Herbal medicine and treatment of diabetes in Africa : an example from Guinea. *Diabetes Metab* 2006, 32(2):171-175.

22. Singh S, Gupta S, Sabir G, Gupta M, Seth P: Database for anti-diabetic plants with clinical/experimental trials. *Bioinformation* 2009, 4(6):263-268.

23. Maritim AC, Sanders RA, Watkins JB 3rd: Diabetes, oxidative stress, and antioxidants: a review. *J Biochem Mol Toxicol* 2003, 17(1):24-38.

24. Kuete V, Efferth T: Cameroonian medicinal plants: pharmacology and derived natural products. *Front Pharmacol* 2010, 1:123.

25. AbouZid S, Elshahaat A, Ali S, Choudhary M: Antioxidant activity of wild plants collected in Beni-Sueif. Upper Egypt. *Drug Discov Ther* 2008, 2(5):286-288. PubMed Abstract

26. Talukder F, Khan K, Uddin R, Jahan N, Alam A: In vitro free radical scavenging and anti-hyperglycemic activities of Achyranthes aspera extract in alloxan-induced diabetic mice. *Drug Discoveries Therapeut* 2012, 6(6):298-305.

27. Houacine CH, Elkhawad AO, Ayoub SMH: A comparative study on the anti-diabetic activity of extracts of some Algerian and Sudanese plants. *J Diabet Endocrin* 2012, 3(3):25-28.

28. Swanston-Flatt SK, Day C, Bailey CJ, Flatt PR: Traditional plant treatments for diabetes. Studies in normal and streptozotocin diabetic mice. *Diabetologia* 1990, 33(8):462-464.

29. Vianna R, Brault A, Martineau LC, Couture R, Arnason JT, Haddad PS: In Vivo Anti-Diabetic Activity of the Ethanolic Crude Extract of *Sorbus decora* C.K.Schneid. (Rosacea): A Medicinal Plant Used by Canadian James Bay Cree Nations to Treat Symptoms Related to Diabetes. *Evidence-Based Complement Altern Med* 2011, 7.

30. Kadan S, Saad B, Sasson Y, Zaid H: *In Vitro* Evaluations of Cytotoxicity of Eight antidiabetic Medicinal Plants and Their Effect on GLUT4 Translocation. *Evidence-Based Complement Altern Med* 2013, 9.

31. Bussmann R, Malca G, Glenn A, Sharon D, Nilsen B, Parris B, Dubose D, Ruiz D, Saleda A, Martinez M, Carillo L, Walker K,

Kuhlman A, Townesmith A: Toxicity of medicinal plants used in traditional medicine in Northern Peru. *J Ethnopharmacol* 2011, 137(1):121.

Chapter 7

Antioxidant Activity of Herbaceous Plant Extracts Protect against Hydrogen Peroxide-induced DNA Damage in Human Lymphocytes

Kuan-Hung Lin[1], Yan-Yin Yang[2], Chi-Ming Yang[3], Meng-Yuan Huang[3], Hsiao-Feng Lo[4], Kuang-Chuan Liu[5], Hwei-Shen Lin[2], and Pi-Yu Chao[6]

[1]Graduate Institute of Biotechnology, Chinese Culture University, Taipei 11114, Taiwan

[2]Graduate Institute of Applied Science of Living, Chinese Culture University, Taipei 11114, Taiwan

[3]Research Center for Biodiversity, Academia Sinica, Nankang, Taipei 11106, Taiwan

[4]Department of Horticulture and Landscape Architecture, National Taiwan University, Taipei 11111, Taiwan

[5]Taoyuan District Agricultural Research and Extension Station, Taoyuan 327 Taiwan

[6]Department of Nutrition and Health Sciences, Chinese Culture University, Taipei 11114, Taiwan

ABSTRACT

Background

Herbaceous plants containing antioxidants can protect against DNA damage. The purpose of this study was to evaluate the antioxidant substances, antioxidant activity, and protection of DNA from oxidative damage in human lymphocytes induced by hydrogen peroxide (H_2O_2). Our methods used acidic methanol and water extractions from six herbaceous plants, including *Bidens alba*(BA), *Lycium chinense* (LC), *Mentha arvensis* (MA), *Plantago asiatica* (PA), *Houttuynia cordata*(HC), and *Centella asiatica* (CA).

Methods

Antioxidant compounds such as flavonol and polyphenol were analyzed. Antioxidant activity was determined by the inhibition percentage of conjugated diene formation in a linoleic acid emulsion system and by trolox-equivalent antioxidant capacity (TEAC) assay. Their antioxidative capacities for protecting human lymphocyte DNA from H_2O_2-induced strand breaks was evaluated by comet assay.

Results

The studied plants were found to be rich in flavonols, especially myricetin in BA, morin in MA, quercetin in HC, and kaemperol in CA. In addition, polyphenol abounded in BA and CA. The best conjugated diene formation inhibition percentage was found in the acidic methanolic extract of PA. Regarding TEAC, the best antioxidant activity was generated from the acidic methanolic extract of HC. Water and acidic methanolic extracts of MA and HC both had better inhibition percentages of tail DNA% and tail moment as compared to the rest of the tested extracts, and significantly suppressed oxidative damage to lymphocyte DNA.

Conclusion

Quercetin and morin are important for preventing peroxidation and oxidative damage to DNA, and the leaves of MA and HC extracts may have excellent potential as functional ingredients representing potential sources of natural antioxidants.

BACKGROUND

Herbaceous plants have a long history of use as medicine, food, and a variety of daily needs. Many epidemiological studies suggest that an increased consumption of several medicinal plants containing antioxidants can protect against DNA damage and carcinogenesis, and often exhibit a wide range of pharmacological activities such as antiflammatory, anti-bacterial, and anti-fungal properties [1]. Flavonoids have strong antioxidant efficiencies and are common in leafy vegetables. Trolox, for example, is a water-soluble derivative of vitamin E that blocks DNA fragmentation in irradiated MOLT-4 cells, a human lymphocytic leukemia line [2]. Hence, a number of phytochemicals commonly used in research have antioxidant activity that can protect cells from reactive oxygen species (ROS)-mediated DNA damage that results in mutation and subsequent carcinogenesis [3,4]. Cao *et al.*[5] indicated that increased consumption of vegetables and fruits increases the plasma antioxidant capacity in humans. Some common vegetables like purple-leaved sweet potato and the outer layers of purple onions abound in quercetin and myricetin, which scavenge 2, 2-diphenyl-1-picrylhydrazyl (DPPH), superoxide, and hydroxyl radicals, and inhibit lipid peroxidation [6]. The search for phytochemicals and dietary compounds with potent antioxidant and otherwise preventive properties continues to be of great importance in the search for remedies against free radical-mediated diseases. There is great interest in the use of potent dietary antioxidants in preventive strategies for applications ranging from the prevention of oxidative reactions in foods and pharmaceuticals to the role of ROS in chronic degenerative diseases [7].

In recent years, increasing attention has been paid by consumers to the health and nutritional benefits of herbaceous plants. Some herbs, such as pilosa beggarticks (*Bidens alba* L. var. *minor*) (BA),

Chinese wolfberry (*Lycium chinense* Mill.) (LC), wild mint or corn mint (*Mentha arvensis* L. var. piperascens Malinv.) (MA), Asiatic plantain (*Plantago asiatica* L.) (PA), heartleaf (*Houttuynia cordata* Thunb.) (HC), and Asiatic centella (*Centella asiatica* L. Urban) (CA) are favored as functional herbals. Some of the health effects of herbaceous plants have been reported to include antioxidation [8-10], anti-inflammation [11], and blood pressure reduction [12]. In animal experiments, Chinese wolfberry, heartleaf, Asiatic plantain, Asiatic centella, and pilosa beggarticks showed special detoxification and anti-inflammatory effects [8, 9, 11, 13, 14]. Particularly, HC, LC, and CA showed antioxidant activities [8, 9]. Asiatic centella increased the activity of antioxidant enzymes such as superoxide dismutase, catalase, and glutathione peroxidase, and enhanced the concentration of vitamin C and vitamin E in new tissues during wound healings [13]. Both HC and BA were reported to have anti-inflammatory functions due to their quercetin and luteolin content [8, 11]. Furthermore, LC and BA can reduce the injury to liver cells from CCl_4 [9,13]. Pilosa beggarticks also functions as an anti-fungal and anti-bacterial agent, and lowers high blood pressure [12]. Several herbs are consumed to protect against common, serious diseases such as cardiovascular and cerebrovascular events, cancer, and other age-related degenerative diseases[15]. These protective effects are considered, in large part, to be related to the various antioxidants contained in them. Evidence that free radicals cause oxidative damage to lipids, proteins, and nucleic acids is overwhelming. Antioxidants, which can inhibit or delay the oxidation of an oxidizer in a chain reaction, would therefore seem to be important in preventing these diseases [16]. Prevention from oxidative stress might be achieved by the uptake of antioxidants. Polyphenols and flavonols can act as antioxidants in two ways: by scavenging free radicals and chelating redox active metal ions (direct antioxidant activity), and by inducing cellular antioxidant defense and repair. These benefits have significantly contributed to their antioxidant activity and have stimulated research into the content, ability, capacity, and function of antioxidant systems in herbaceous plants. Polyphenolic and flavonol substances are the most common compounds in herbs having strong antioxidant activity [6]. Previously, we also demonstrated that purple-leaved sweet potato exhibits free radical scavenging and has high polyphenolic content [17]. Although a variety of medicinal herbs are known to be potent sources of polyphenolic and flavonol compounds, studies that isolate polyphenols, evaluate their antioxidative effects, and determine their efficacy or ability to prevent oxidative damage to DNA are either scarce or little known. The bioactive components of these herbal plants might be responsible for anti-cancer effects through growth inhibition and apoptosis in human chronic myeloid leukemia K562 cells [18]. The objective of this study was to isolate, identify, and evaluate the antioxidant components, antioxidant activity, and extent to which methanolic acid hydrolysates and water extracts of six herbaceous plants could protect DNA in human lymphocytes from oxidative damage induced by H_2O_2. Our study explores the relationships between the composition and content of flavonols and polyphenol having antioxidant efficiency, and the prevention of DNA oxidative damage afforded by the herbaceous plants.

METHODS

Chemicals and Reagent

Methanol, ethanol, hydrochloric acid, di-sodium hydrogen phosphate, potassium dihydrogen phosphate, formic acid, sodium chloride (NaCl), potassium chloride (KCl), Tris–HCl, Tris (hydroxymethyl) aminomethane (Tris base), dimethyl sulfoxide (DMSO), ethylenediamine tetraacetic acid (EDTA), Trolox, and butylated hydroxyltoluene were purchased from Merck (Darmstadt, Germany). Linoleic acid, d-glucose, calcium chloride dihydrate, sodium lauryl sarcosinate, gallic acid, 2,2-azino-bis-(3-ethylbenzothiazoline-6-sulfonicacid) (ABTS), peroxidase, H_2O_2, sodium carbonate (Na_2CO_3), tetrazolium/formazan, Folin-Ciocalteau reagent, and ethidium bromide were procured from Sigma Chemical (St Louis, MO, USA). Myricetin, morin, quercetin, kaempferol, cynidin, and malvidin were obtained from ROTH (Rheinzabern, Denmark). Ficoll-Paque was acquired from Amersham Biosciences (Uppsala, Sweden). Low-melting gel agrose and Triton X-100 were purchased from BDH (Poole, England). Normal-melting gel agarose was purchased from Pantech Instruments (Darmstadt, Germany). AIM V serum-free lymphocyte medium was purchased from Gibco Invitrogen (Carlsbad, CA, USA).

Herbaceous Plants

The tested plants were *Bidens alba* L. var. *minor, Lycium chinense* Mill., *Mentha arvensis* L. var. piperascens Malinv., *Plantago asiatica* L., *Houttuyni acordata* Thunb., and *Centella asiatica* L. Urban. These were generously provided by Dr. Kuang-Chuan Liu, Taoyuan District Agricultural Research and Extension Station Council of Agriculture, Executive Yuan, Taiwan.

Preparation of Plant Extracts

The plants were weighed, lyophilized, and ground to powder. Each lyophilized powder was extracted by distilled deionized (dd) H_2O. The extraction mixture was then heated to 90°C in a steam bath and refluxed for 2 h, allowed to cool in a refrigerator, sonicated for 5 min, and diluted to 50 mL with ddH$_2$O to prepare the final extract. These water extracts were ready for the comet assay. For high-performance liquid chromatography (HPLC), only the edible portions of plants were weighed, lyophilized, and ground into powder. Lyophilized vegetable powders were prepared according to Justesen et al. [19] with modifications as follows: 10 ml of 62.5% aqueous methanol containing butylated hydroxyltoluene (2 g/L) were added to 1.25 g of lyophilized samples, followed by adding 5 mL of 6 M HCl to bring total volume up to 12.5 mL. The final mixture consisted of 1.2 M HCl in 50% aqueous methanol. The extraction mixture was thereafter heated to 90°C in a steam bath and refluxed for 2 h, allowed to cool in a refrigerator, sonicated for 5 min, and diluted to 50 mL with methanol to form the final extract. The acid hydrolysates methanolic extract was ready for high-performance liquid chromatography (HPLC), inhibition of conjugated diene formation in the linoleic acid assay, TEAC assay, and comet assay.

Polyphenol Assay

Polyphenol content was determined according to the method of Taga et al. [20]. Briefly, standard gallic acid and an aliquot of methanolic extract were diluted with an ethanol/water (60:40, v/v) solution containing 0.3% HCl. Two mL of 2% Na_2CO_3 was mixed into each sample of 100 µL and allowed to equilibrate for 2 min before adding 50% Folin-Ciocalteau reagent. Absorbance at 750 nm was measured at room temperature. The standard curve of gallic acid was used to calculate polyphenol levels.

Flavonols Analysis by HPLC

One mL of acid hydrolysates methanolic extract was filtered through a 0.45 µm filter prior to 20 µL being injected into the HPLC. Samples were analyzed with a SpectraSYSTEMUV6000LP Photodiode Array Detection System (Thermo Separation Products, San Jose, USA) and an ODS column (250 × 4.6 mm, 5 µm; YMC, Kyoto, Japan). The mobile phase consisted of methanol–water (30:70, v/v) with 1% formic acid and 100% methanol. The gradient was 25 - 74% methanol in 40 min at a flow rate of 0.75 mL/min. Spectra were recorded at 365 nm for flavonols [19].

Inhibition of Conjugated Diene Formation in Linoleic Acid Emulsion Autoxidation System

The inhibition of conjugated diene formation was determined according to Mitsuda et al. [21]. Briefly, an aliquot of 0.1 mL of diluted plant acidic methanolic extract or blank was added to 2 mL of 10 mM linoleic acidemulsion (pH 6.6), mixed well, and incubated at 37°C for 15 h. A sample of 0.2 mL for 0 and 15 h incubation periods were mixed with 7 mL of 80% methanol, followed by measuring the absorbance at 234 nm.

Trolox Equivalent Antioxidant Capacity (TEAC) Analysis

The total antioxidant capacity of hydrophilic and lipophilic antioxidants was determined using the horseradish peroxidase catalyzed oxidation of 2,2-azino-bis-(3-ethylbenzothiazoline-6-sulfonicacid) (ABTS) [22]. The reaction mixture contained 0.5 mL of 1000 µM ABTS (in ddH_2O) and 3.5 mL of 100 µM H_2O_2 (in ddH_2O). The reaction was started by adding 0.5 mL of 44 U/mL peroxidase (in 0.1 M PBS). After 1 h, 0.05

mL of plant acidic methanolic extracts were added to the mixture. After 5 min, absorbance was measured at 730 nm. Trolox (TR) was used as a standard, and the total antioxidant capacity of plant extracts were measured as mM TR equivalent.

Isolated Human Peripheral Blood Lymphocytes

Fasting blood samples were obtained from six donors, including four male and two female healthy non-smokers, 24–48 years old. Fresh venous blood (20–30 mL) was collected in lithium heparin tubes (Becton- Dickinson) from volunteers, and lymphocytes were isolated using a separation solution kit supplemented with Ficoll-Paque Plus lymphocyte isolation sterile solution (Pharmacia Biotech, Sweden) [23]. Cells were harvested within 1 day of taking the blood samples and cultured with AIM V serum-free lymphocyte medium (Gibco Invitrogen, USA) in a humidified atmosphere of 5% CO_2 in air at 37°C for 24 h.

Cell Viability Testing

After culturing, lymphocytes were exposed to each of six different plant acidic methanolic and water extracts. Each lymphocyte was treated with three concentrations of plant acidic methanolic and water extracts (25, 50, and 100 μg/mL) for 30 min at 37°C. DNA damage was induced by exposing lymphocytes to H_2O_2 (10 μM) for 5 min on ice to minimize the possibility of cellular DNA repair after H_2O_2 injury. Cells were centrifuged (100 g for 10 min), washed, and re-suspended in the same medium as the comet assay. All experiments were carried out in triplicate. Cell viability was tested using the tetrazolium/formazan (MTT) assay [24] both prior to and after treatment with plant extracts or H_2O_2.

DNA Single Strand Break Damage Estimation Using the Comet Assay

The standard comet assay was performed as described in Szeto et al. [3], with acidic methanolic and water extracts from these six herbal plants being used for this study. Cultured lymphocytes (10^5 cells/

mL) were embedded in 75 μL of 1% low-melting-point agarose on a microscope slide (precoated with agarose) at 37°C. The gel was allowed to set at 4°C, and cells were lysed for a period of at least 2 h in lysis buffer at 4°C. Cells were then alkaline-unwound, following which electrophoresis was carried out using the electrophoresis buffer at 4°C for 15 min at 25 V with the current adjusted to 300 mA. All steps were conducted under dim light to prevent the occurrence of additional DNA damage. Following electrophoresis, slides were neutralized with neutralization buffer and stained with ethidium bromide. The comet-like images resulting from the extension of DNA were scored as a reflection of the single strand breaks under a fluorescence microscope (Zeiss-Axiovert 100, Zeiss, Germany). Triplicate slides were prepared for each experimental point sample, and 50 comet-like images selected at random per slide were evaluated to determine average DNA damage values. A computerized image analysis system (VisCOMET 1.6, Impuls GmbH, Germany) was employed to determine various comet parameters, and used to analyze DNA damage by tail DNA% [(total brightness of tail area / total brightness of total area) × 100%] and tail moment (tail length × tail DNA%). Inhibition percentage of tail DNA% and tail moment were calculated relative to the 10 μM H_2O_2 treated group.

Statistical Analysis

Data were analyzed by one-way analysis of variance (ANOVA), and the significance between means by the least significant difference (LSD) test. Pearson's linear correlation was also determined. Means of three replicates were reported.

RESULTS

Antioxidant Composition and Antioxidant Activity

Table 1 documents the content of polyphenol in the leaves of tested plants. Polyphenols were significantly abundant in both BA (32.90

mg gallic acid/g DW) and CA (32.03 mg gallic acid/g) compared to other plants. Table 2 presents varied amounts of flavonols ranging from 53.33 to 3200 µg/g DW in the acidic methanolic extract of the studied plants. BA and CA were also rich in myricetin, at levels of 1133.33 and 960.00 µg/g DW, respectively. Morin was present only in MA, CA, and BA plants at a level of 2000.00, 600.00, and 573.33 µg/g DW. Quercetin was abundant in HC (3200.00 µg/g DW), while CA followed at a level of 533.33 µg/g DW. Kaempferol was abundant in CA at a level of 853.33 µg/g DW, but LC and PA did not contain any kaempferol at all. Thus, these species displayed variations in their polyphenol and flavonol levels.

Table 1: The content of polyphenol in tested herbaceous plants

Sample	Ployphenol (mg gallic acid/g DW)
BA	32.90[a]
LC	25.31[b]
MA	21.24[c]
PA	24.31[b]
HC	19.82[d]
CA	32.03[a]

Means with different superscripts ([a-d]) are significantly different, $p < 0.05$.

DW: dry weight.

Lin *et al.*

Lin *et al.* *BMC Research Notes* 2013 6:490, doi:10.1186/1756-0500-6-490

Table 2: The content of flavonols in acidic methanolic extracts of tested herbaceous plants

Flavonols (µg/g DW)			
Sample	Myricetin	Morin	Quercetin	Kaempferol
BA	1133.33[a]	573.33[c]	93.33[d]	66.67[c]
LC	320.00[d]	N.D.	N.D.	N.D.
MA	480.00[c]	2000.00[a]	53.33[e]	333.33[b]

PA	253.33[e]	N.D.	213.33[c]	N.D.
HC	146.67[f]	N.D.	3200.00[a]	53.33[c]
CA	960.00[b]	600.00[b]	533.33[b]	853.33[a]

Means within a column with different superscripts ([a-f]) are significantly different, $p < 0.05$. N.D.: not detectable.

Lin *et al.*

Lin *et al. BMC Research Notes* 2013 **6**:490, doi:10.1186/1756-0500-6-490

The inhibition of linoleic acid peroxidation was observed to be significantly higher in PA and BA at both 25 and 50 µg/mL of plant extracts (Table 3). Furthermore, significantly higher percentages of conjugated diene inhibition were detected in PA (79.31) and CA (77.61) compared to MA (70.31) at 100 µg/mL of the extract. Hence, each species showed significant differences in inhibition percentages of conjugated dienes at various extract concentrations.

Table 3: Inhibition percentage of conjugated diene formation in the linoleic acid emulsion autoxidation system treated with various concentrations of methanolic acid hydrolysates of herbaceous plants

Inhibition percentage			
Sample	25 µg/mL	50 µg/mL	100 µg/mL
BA	49.14[ab]	55.71[b]	73.72[b]
LC	40.14[c]	52.31[c]	74.69[b]
MA	46.23[b]	56.20[b]	70.31[c]
PA	54.50[a]	68.37[a]	79.31[a]
HC	45.49[b]	59.61[b]	72.99[b]
CA	41.11[c]	58.63[b]	77.61[a]

Means within a column with different superscripts [a-c] are significantly different, $p < 0.05$.

Lin *et al.*

Lin *et al.* BMC Research Notes 2013 **6**:490, doi: 10.1186/1756-0500-6-490

Plant extracts from the six species showed antioxidant activities, proving their capacity to scavenge the ABTS radical-cation. The antioxidant activity in methanolic acid hydrolysate extracts of leaf tissues of studied species were expressed in Trolox Equivalent Antioxidant Capacity (TEAC) (Table 4). HC showed a significantly higher TEAC value (231.16 mM) than other species.

Table 4: The TEAC values of acidic methanolic extracts in the investigated herbaceous plants

Samples (100 µg/mL)	TEAC (mM Trolox)
BA	184.61[b]
LC	117.44[e]
MA	146.44[c]
PA	142.49[c]
HC	231.16[a]
CA	132.80[d]

Means within a column with different superscripts [a-e] are significantly different, $p < 0.05$.

Lin *et al.*

Lin *et al.* BMC Research Notes 2013 6:490, doi:10.1186/1756-0500-6-490

Effects of Acidic Methanolic and Water Extracts from Herbaceous Plants on H_2O_2-Induced DNA Damage to Lymphocytes

Lymphocytes were exposed to each of three different herbal extracts at three concentrations (25, 50, and 100 µg/mL) for 30 min at 37°C. DNA damage was induced by exposing lymphocytes to H_2O_2 (10 µM)

for 5 min on ice. At two lower levels, no extracts were cytotoxic at the concentrations used, with > 98% of cells remaining viable [25]. Therefore, concentrations only at 25 and 50 µg/mL were chosen for the comet assay. The comet assay was performed to determine the DNA damaging activity of the plants as it is a sensitive method for monitoring single strand DNA breaks at the single cell level. Any DNA damage is represented as tail DNA% and tail moment. The effects of pretreatment of the six tested extracts on 10 µM H_2O_2-induced DNA oxidative damage in human lymphocytes are presented in Figure 1. Tail DNA% demonstrated that MA had a significantly greater level of protection against H_2O_2 exposure than lymphocytes that were exposed to other tested compounds at two doses (25 and 50 µg/mL) (Figure 1A). The maximum protective effect of lymphocyte pretreatment was observed with pretreatment by 25 µg/mL MA, exhibiting 12.43% of tail DNA% compared to the rest of treated samples. Furthermore, at lower concentrations, all tested samples had lower tail DNA%, indicating better inhibition efficacies. The MA extract at the 50 µg/mL was significantly lower than the rest of treated samples, except for HC extract. Tested plants showed at least 707.53 and 1040.63 of tail moment in HC extract at 25 and 50 µg/mL levels compared to the rest of the acidic methanolic extract samples (Figure 1B).

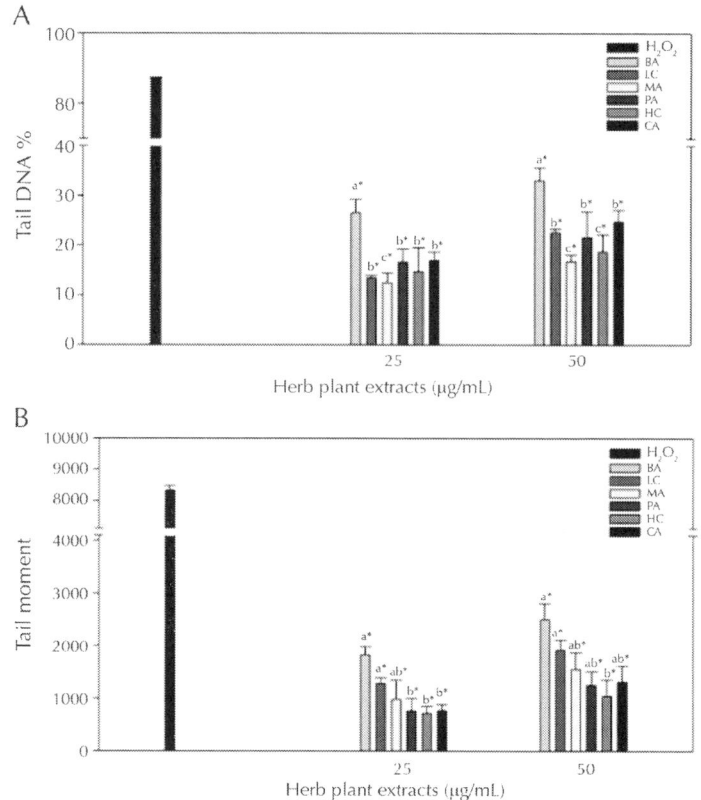

Figure 1: Effects of various acidic methanolic extracts from six herbaceous plants on H2O2-induced DNA damage to lymphocytes. Tail DNA% (A) and tail moment (B) were measured after exposure to tested compounds at 25 and 50 µg/mL of extract. ▢, BA;■, LC; ▢, MA; ■, PA; ▨, HC; ■, CA. Values with different letters differ significantly with regard to oxidative damage when comparing different plant extracts; *$p < 0.05$ refers to differences in oxidative damage as compared with 10 µM H_2O_2-alone (■) treatment.

HC had the lowest % tail DNA at 11.14% in 25 µg/mL of water extract (Figure 2A). Both HC (18.36%) and MA (18.25%) extracts at 50 µg/mL had lowest % tail DNA compared to the rest of the water extract of samples. Moreover, HC also had a significantly lower tail moment (1255.40 ~ 1826.10) than the rest of the water extracts at the same doses (Figure 2B). Hence, the DNA damage induced by H_2O_2 was significantly high as compared to the treated extracts, which had 87.26 in tail DNA% and 8328.84 in tail moment.

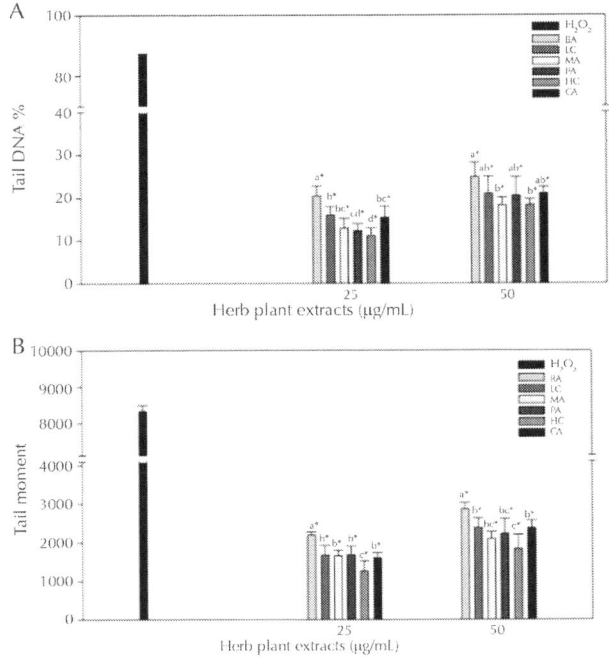

Figure 2: Effects of various water extracts from six herbaceous plants on H2O2-induced DNA damage to lymphocytes. Tail DNA%(A) and tail moment (B) were measured after exposure to tested compounds at 25 and 50 µg/mL of extract. ▢, BA; ▦, LC; ▢▦, PA; ▦, HC; ■, CA. Values with different letters differ significantly with regard to oxidative damage when comparing different plant extracts; *$p < 0.05$ refers to differences in oxidative damage as compared with 10 µM H_2O_2-alone (■) treatment.

DISCUSSION

Antioxidant Composition and Antioxidant Activity

Plant leaves are rich in flavonols and other pigments. BA and CA plants contain higher polyphenol levels than the other plants tested (Table 1). Antioxidant activities are known to increase proportionally to the polyphenol content, mainly due to their redox properties [1].

Among the diverse roles of polyphenols, they protect cell constituents against destructive oxidative damage, thus limiting the risk of various degenerative diseases associated with oxidative stress and tending to be potent free radical scavengers. Their ability to act as antioxidants depends on their chemical structure and ability to donate/accept electrons, thus delocalizing the unpaired electron within the aromatic structure [26]. Phenolic compounds are known as radical scavengers or radical-chain breakers, and they strongly eliminate oxidative free radicals. Quercetin and morin are the principal flavonol constituents in HC and MA plants, respectively (Table 2). These antioxidant compounds may account for the high antioxidant power of the plants in the present study. Quercetin, kaempferol, morin, and myricetin are the most common flavonols, and are the most widely distributed flavonoids in plant leaves. Quercetin, the most abundant flavonoid in the human diet, is an excellent free radical scavenging antioxidant [27]. Polyphenol and flavonol contents found in the extracted plants (Tables 1, 2) were much lower than those in our previous study where purple-leaved sweet potato appeared to have higher contents [28]. A possible reason is the usage of different extraction methods. In fact, different results were obtained from the water and acidic methanolic extracts, and especially from the water extracts. The antioxidant composition and activities of herbal plants cannot be evaluated by a single method due to the complex nature of plants, in which pigments and phytochemicals have specific functions. Therefore, several methods should be employed to evaluate the total antioxidant effects of any plant. Antioxidant compounds presented in plant extracts are therefore multi-functional and their activities and mechanisms of action would largely depend on the composition and conditions of the test system.

Compared to the inhibition percentage of conjugated diene formation in the linoleic acid emulsion autoxidation system of tested samples, PA exhibited relatively higher effectiveness than the others at all extract concentrations (Table 3). The tested vegetables showed >70% inhibition of linoleic acid peroxidation in 100 µg/mL extracts, and PA in particular exhibited the highest inhibition of linoleic acid peroxidation, up to 79.31%. Therefore, all tested plants were effective inhibitors and exhibited better inhibition efficacy at higher concentrations. Previously, we demonstrated that water and methanolic extracts from PA both had higher antioxidant activity, and that the antioxidant activity of PA was equivalent to 10^{-4} M of Trolox in preventing conjugated diene formation during linolic acid peroxidation at 62.5 µg/mL of methanolic extract [29]. The polyphenol content of methanolic extracts was significantly correlated with the delay of the lag phase of low-density lipoprotein (LDL) treated with methanolic extracts. Moreover, the polyphenol content of the methanolic extract of herbaceous plants was significantly correlated with scavenging DPPH radical activity and ferric reducing power [29].

We measured the direct antioxidant activity of acidic methanolic extracts by TEAC assay, reflecting the major mechanisms of antioxidant action for evaluating their relevance to cell protection (Table 3). Jastrzebski et al.[30] reported that prolipid, a mixture of herbs used as a plasma lipid lowering medicine, had strong antioxidant activity. The correlation coefficients between the polyphenols, flavonoids, and TEAC of prolipid water extracts were 0.97 and 0.90, respectively. They concluded that the content of polyphenol in prolipid was the main contributors to the overall antioxidant activity of prolipids. The antioxidant activity of leaf extracts from CA was found to have a direct linear relationship between total phenolic content and total antioxidant activity, indicating that phenolic compounds might be the major contributors to the antioxidant activities of CA extracts [31]. Chung et al.[29] reported that PA, BA, CA, Curled Spearmint, MA, and Mesona had higher total phenolic contents compared to LC and Taiwan lily, and that CA and PA had higher antioxidant activity. In this study, we found that HC and CA contained abundant quercetin while MA and CA were rich in morin and kaempferol, respectively. Additionally, BA and CA had significantly higher levels of myricetin than other tested samples (Table 2). These different pigments may exhibit effective antioxidant activity alone or synergistically, and are a likely cause of cultivar differences. Wang et al. [32] demonstrated that the H donation potential was quercetin > myricetin > morin > kaempferol, indicating that the presence of a 3',4'-catechol moiety in the B ring correlated with high activity. Moreover, the structural peculiarity of di-OH in the B ring obviously rendered quercetin and morin more potent as ROS inhibitors than myricetin and kaempferol, which have tri- and mono-OH in the B ring, respectively. The unclear relationship between antioxidant activity and flavonol extracts indicates that the structure prerequisite to reinforce free radical scavenging activity may vary with the type of free radical. The synergisms among antioxidants make antioxidant activity dependent not only on the concentration, but also might be due to their structures and interactions among antioxidants [33]. The accumulation of flavonoid metabolites in the appropriate target site is probably required to exert their antioxidant activity. The polyphenol-rich plant extracts exhibited distinct cell-free antioxidant activity (TEAC) according to their levels of polyphenol and flavonols, with distinct antioxidant activity strongly accounting for the antioxidant activity of the extracts. HC plants containing 3200 µg/g DW quercetin (Table 2) exhibited the highest TEAC value (231.16 mM) within the tested extracts (Table 4).

Estimation of DNA Single Strand Break Damage from Exposure to Acidic Methanolic and Water Extracts

Quercetin was found to protect against H_2O_2-induced DNA damage in human lymphocytes at 10 µM [34] and at 3.1 to 25 µM [35]. However, it was found to induce DNA damage in human lymphocytes at higher concentrations, such as 100 µM or above [34]. Similarly, myricetin was also found to decrease oxidant-induced DNA damage at 100 µM, although α–tocopherol and β–carotene did not behave similarly. This might be due to the dihydroxy structure of quercetin and myricetin being essential for protecting DNA against hydrogen peroxide [34]. No such hydroxyl groups are present in the tocopherol molecule. This may reflect structure/activity relationships or the localization of the antioxidant relative to free radical generation within cells. Noroozi et al. [36] demonstrated that, in addition to quercetin, kaempferol could also inhibit H_2O_2-induced DNA strand breaks in human lymphocytes. Zhu and Loft [37] reported that aqueous extracts of cooked and autolysed Brussels sprouts decreased DNA strand breaks in human lymphocytes, with the maximum inhibition being 38 and 39% at cooked and autolysed extract levels of 10 µg/mL and 5 µg/mL, respectively, with the inhibition effect decreasing at increasing concentrations up to 100 µg/mL. Quercetin-rich onions showed increased resistance of lymphocytic DNA to ex vivo-induced oxidation [15]. In addition, several types of natural antioxidants, including flavonols and polyphenolic compounds, inhibit adhesion molecule expression and the adhesion of monocytes to endothelial cells, and also suppress cell inflammation, transformation, proliferation, survival, invasion, and angiogenesis [38-40]. Free radicals induce cellular damage and are involved in several human diseases such as cancer, atherosclerosis, and inflammatory disorders, and polyphenols tend to reduce mutagenic activity and oxygen-free radicals [41]. Since the initiation and progress of carcinogenesis involves mutations of DNA, the chemical alteration of DNA bases is believed to be a crucial factor. As a consequence of increased oxidative stress, DNA oxidation damage can occur with ROS, leading to mispairing of DNA bases or DNA strand breaks. ROS are generated endogenously from cellular metabolism and inflammatory responses or by exposure to exogenous agents such as ionizing radiation and xenobiotics [42].

In our study, the inhibition percentages of tested plants ranged from 74.51% (BA) to 91.45% (MA) with acidic methanolic extract concentrations at 25 µg/mL (Figure 1A). MA plants had a value of 985.73 (91.95% inhibition percentage) for tail moment at 25 µg/mL of acidic methanolic extracts (Figure 1B). The results in inhibition percentage of tail DNA% were not similar to the results in inhibition percentage of tail moment among treated samples. The MA plant extract was most effective against DNA single strand breaks in tail DNA%, while HC plant extract was most effective against DNA single strand breaks in tail moment (Figure 1A and 1B). In addition, HC plant water extracts exhibited 11.14% tail DNA% (Figure 2A) and 1255.40 (92.19% inhibition percentage) tail moment at the 25 µg/mL dose (Figure 2B). The inhibition percentage of tail DNA% results was similar to the results of the inhibition percentage of tail moment among treated samples. HC plant extracts not only had the highest Trolox equivalent (Table 4), but were also the most effective against DNA single strand breaks induced by H_2O_2 in human lymphocytes (Figure 1), indicating that it contains polyphenol (19.82 mg gallic acid/g DW), myricetin (146.67 µg/g DW), quercetin (3200.00 µg/g DW), and kaempferol (53.33 µg/g DW) (Tables 1 and 2). To some extent, the observed efficacy of the extracts against DNA damage can be attributed to specific flavonol constituents. The high levels of quercetin and morin are believed to account for the high DNA protective potential of HC and MA since quercetin has also been identified as an efficient reducer of DNA damage in Caco-2 cells [43]. Morin from *Psidium guajava* was effective in increasing cell viability, decreasing ROS levels, and preventing DNA fragmentation upon exposure to high glucose levels in primary rat hepatocyte cultures [44]. The antioxidant activity of polyphenolic compounds in different species showed higher polyphenolic content and antioxidant activity in all species, demonstrating that the tested species are a potent source of novel bioactive compounds with a wide range of medicinal properties. In particular, they have significant free radical scavenging activity. Our present study demonstrates that, among the six investigated species, the higher content of polyphenols, flavonols, and antioxidant properties in HC and MA plants may be the reason for their wide medicinal use. Both species can be used as potent medicinal herbs for novel bioactive compounds with high free radical scavenging activity, and extracts of these plants may been attractive alternative for managing oxidative stress-induced liver injury and drug-induced gastric ulcer [45,46]. Recently, Gargouri et al. [47] demonstrated that quercetin could protect against dimethoate-induced oxidative stress by decreasing lipid peroxidation and protein oxidation, and increasing superoxide dismutase and catalase activities in human lymphocytes. The herbaceous plant extracts in our study may increase antioxidant enzyme activities to protect against H_2O_2-induced DNA damage in human lymphocytes.

CONCLUSIONS

Polyphenol-rich extracts from the tested plants effectively diminish DNA oxidation damage. This preventive effectiveness is attributable to the induction of cellular defenses rather than the radical scavenging activity of polyphenol and flavonols, and might well contribute to the reported health benefits of herbals. The contents of these bioactive compounds in MA and HC extracts can explain their antioxidant activity, and there exists a relationship between the content of polyphenol and flavonol to antioxidant activity. This is the first report suggestion that MA and HC plants have abundant antioxidants with strong antioxidant activity, and consequently can protect DNA in lymphocytes from oxidative damage.

AUTHORS' CONTRIBUTIONS

KHL prepared the extracts and carried out all the experimental process. PYC designed the current project, supervised the work and wrote the manuscript. YYY worked closely with KCL and MYH in the laboratory to carry out the experiments. HFL and HSL evaluated the data and edited the manuscript. CMY participated in statistical analysis. All the authors read and approved the final manuscript.

REFERENCES

1.   Rasineni GK, Siddavattam D, Reddy AR: Free radical quenching activity and polyphenols in three species of Coleus. *J Med Plants Res* 2008, 2:285-291.

2.   McClain DE, Kalinich JF, Ramakrishnan N: Trolox Inhibits apoptosis in irradiated MOLT-4 lymphocytes. *FASEB J* 1995, 9:1345 1354.

3.   Szeto YT, Collins AR, Benzie IF: Effects of dietary antioxidants on DNA damage in lysed cells using a modified comet assay procedure. *Mutat Res* 2002, 500:31-38

4.   Lazzé MC, Pizzala R, Savio M, Stivala LA, Prosperi E, Bianchi L: Anthocyanins protect against DNA damage induced by tertbutyl-hydroperoxide in rat smooth muscle and hepatoma cells. *Mutat Res* 2003, 535:103-115.

5.   Cao G, Sofic E, Prior RL: Antioxidant capacity of tea and common vegetables. *J Agric Food Chem* 1996, 44:3426-3431

6.   Chu YH, Chang CL, Hsu HF: Flavonoid content of several vegetables and their antioxidant activity. *J Sci Food Agr* 2000, 80:561-566.

7.   Farombi EO, Hansen M, Ravn-Haren G, Moller P, Dragsted LO: Commonly consumed and naturally occurring dietary substances affect biomarkers of oxidative stress and DNA damage in healthy rats. *Food Chem Toxicol* 2004, 42:1315-1322.

8.   Taguchi K, Hagiwara Y, Kajiyama K, Suzuki Y: Pharmacological studies of Houttuyniae herba: the anti-inflammatory effect quercitrin. *Yakugaku Zasshi* 1993, 113:327-333

9.   Kim HP, Kim SY, Lee EJ, Kim YC: Zeaxanthin dipalmitate from Lycium chinese has hepatoprotective activity. *Res Commun Mol Pathol Pharmacol* 1997, 97:301-314.

10.   Chen YY, Liu JF, Chen CM, Chao PY, Chang TJ: A study of the antioxidative and antimutagenic effects of Houttuynia cordata

Thunb using an oxidized frying oil-fed model. *J Nutr Sci Vitaminol* 2003, 49:327-333.

11. Geissberger P, Sequin U: Constituents of *Bidens pilosa* L.: Do the components found so far explain the use of this plant in traditional medicine? *Acta Trop* 1991, 48:251-261.

12. Dimo T, Nguelefack TB, Kamtchouing P, Dongo E, Rakotonirina A, Rakotonirina SV:Hyperotensive effects of a methanol extract of Bidens pilosa Linn on hypertensive rats. *C R Acad Sci III* 1999, 322:323-329.

13. Chin HW, Lin CC, Tang KS: The hepatoprotective effects of Taiwan folk medicine ham-hong-chho in rats. *Am J Chin Med* 1996, 24:231-40.

14. Chen YY, Chen CM, Chao PY, Chang TJ, Liu JF: Effects of frying oil and Houttuynia cordata thunb on xenobiotic-metabolizing enzyme system of rodents. *World J Gastroenterol* 2005, 11:389-392.

15. Scalbert A, Johanson IT, Saltmarsh M: Polyphenols: antioxidants and beyond. *Am J Clin Nutr* 2005, 81:215-217.

16. Yen GC, Chuang DY: Antioxidant properties of water extracts from Cassia tora L. in relation to the degree of roasting. *J Agric Food Chem* 2000, 48:2760-2765.

17. Lin KH, Chao PY, Yang CM, Cheng WC, Lo HF, Chang TR: The effects of flooding and drought stresses on the antioxidant constituents in sweet potato leaves. *Bot Stud* 2006, 47:417-426.

18. Liu YL, Tang LH, Liang ZQ, You BG, Yang SL: Growth inhibitory and apoptosis inducing by effects of total flavonoids from Lysimachia clethroides Duby in human chronic myeloid leukemia K562 cells. *J Ethnopharmacol* 2010, 131:1-9.

19. Justesen U, Knuthsen P, Leth T: Quantitative analysis of flavonols, flavones, and flavanones in fruits, vegetables and beverages by high-performance liquid chromatography with photo-diode array and mass spectrometric. *J Chromatogr* 1998, 799:101-110.

20. Taga MS, Miller EE, Pratt DE: Chia seeds as asource of natural lipid antioxidants.*J Am Oil Chem Soc* 1984, 61:928-931.

21. Mitsuda H, Yasumodo K, Iwami K: Antioxidative Action of indole compounds during the autoxidation of linoleic acid. *Eiyo to Shokuryo* 1966, 19:210-214.

22. Re R, Pellegrini N, Evans C: Antioxidant activity applying an improved ABTS radical cation decolorization assay. *Free Rad Biol Med* 1999, 26:1231-1237.

23. Cole J, Green MHL, James SE, Henderson L, Cole H: A further assessment of factors influencing measurements of thioguanine-resistant mutant frequency in circulating T-lymphocytes. *Great Brit Mut Res* 1988, 204:493-507.

24. Cory AH, Owen TC, Barltrop JA, Cory JG: Use of an aqueous soluble tetrazolium/formazan assay for cell growth assays in culture. *Cancer Commun* 1991, 3:207-212

25. Yang YY: *The antioxidative capacity in herb plant extracts and their protection role in DNA oxidative damage of lymphocyte.* Taipei, Taiwan: Chinese Culture University; 2004. [*M.S. Thesis*]

26. Ross JA, Kasum CM: Dietary flavonoids: bioavailability, metabolic effects, and safety. *Annu Rev Nutr* 2002, 22:19-34

27. Villano D, Fernandez-Pachon S, Troncoso AM, Garcia-Parrilla MC: Comparison of antioxidant activity of wine phenolic compounds and metabolites in vitro.*Anal Chim Acta* 2005, 538:391-398

28. Tang SC, Lo HF, Lin KH, Cheng TJ, Yang CM, Chao PY: The antioxidant capacity of extracts from Taiwan indigenous purple-leaved vegetables. *J Taiwan Soc Hort Sci* 2013, 59(1):43-57.

29. Chung AL, Lo H-F, Lin KH, Liu KL, Yang CM, Chao PY: Study on the Antioxidant Activity in Herb Plant Extracts. *J Taiwan Soc Hort Sci* 2013, 59(2):139-152.

30. Jastrzebski Z, Tashma Z, Katrich E, Gorinstein S: Biochemical characteristics of the herb mixture Prolipid as a plant food supplement and medicinal remedy. *Plant Foods Hum Nutr* 2007, 62:145-150.

31. Zaniol MK, Hamid A, Yusof S, Muse R: Antioxidative activity and total phenolic compounds of leaf, root and petiole of four accessions of Centell aasiatica (L). *Urban. Food Chem* 2003, 81:575-581.

32. Wang L, Tu YC, Lian TW, Hing JT, Yen JH, Wu MJ: Distinctive antioxidant and anti-inflammatory effects of flavonols. *J Agric Food Chem* 2006, 54:9798-804

33. Vanderjagt TJ, Ghattas R, Vanderjagt DJ, Glew RH: Comparison of the total antioxidant content of 30 widely used medicinal plants of New Mexico. *Life Sci* 2002, 70:1035-1040

34. Duthie SJ, Collins AR, Duthie GG, Dobson VL: Quercetin and myricetin protect against hydrogen peroxide-induced DNA damage (strand breaks and oxidized pyrimidines) in human lymphocytes. *Mut Res* 1997, 393:223-231.

35. Liu GA, Zheng RL: Protection against damaged DNA in the single cell by polyphenols. *Pharmazie* 2002, 57:852-854

36. Noroozi M, Angerson WJ, Lean ME: Effects of flavonoids and vitamin c on oxidative DNA damage to human lymphocytes. *Am Soc Clin Nutr* 1998, 67:1210-1218.

37. Zhu CY, Loft S: Effects of Brussels sprouts extracts on hydrogen peroxide-induced DNA strand breaks in human lymphocytes. *Food Chem Toxicol* 2001, 39:1191-1197.

38. Moon MK, Lee YJ, Kim JS, Kang DG, Lee HS: Effect of cafeic acid on tumor necrosis factor-alpha-induced vascular inflammation in human umbilical vein endothelial cells. *Biol Pharm Bull* 2009, 32:1371-1377.

39. Li F, Li C, Zhang H, Lu Z, Li Z, You Q, Lu N, Guo Q: A novel flavonoid derivative, inhibits migration and invasion of human breast cancer cells. *Toxico Appl Pharma* 2012, 261:217-226.

40. Chao PY, Huang YP, Hsieh WB: Inhibitive effect of purple sweet potato leaf extract and its components on cell adhesion and inflammatory response in human aortic endothelial cells. *Cell Adh Migr* 2013, 7:237-245.

41. Aviram M: Review of human studies on oxidative damage and antioxidant protection related to cardiovascular diseases. *Free Rad Res* 2000, 33:S85-S87.

42. Bellion P, Digles J, Will F, Janzowski C: Polyphenolic apple extracts: effects of raw material and production method on antioxidant effectiveness and reduction of DNA damage in Caco-2 cell. *J Agr Food Chem* 2010, 58:6636-6642.

43. Schaefer S, Baum M, Eisenbrand G, Dietrich H, Will F, Janzowski C: Polyphenolic apple juice extracts and their major constituents reduce oxidative damage in human colon cell lines. *Mol Nutr Food Res* 2006, 50:24-33.

44. Kapoor R, Kakkar P: Protective role of morin, a flavonoid, against high glucose induced oxidative stress mediated apoptosis in primary rat hepatocytes. *Plos One* 2012, 7(8):e41663. doi:10.1371/journal.pone.0041663

45. Tian L, Shi X, Zhu J, Ma R, Yang X: Chemical composition and hepatoprotective effects of polyphenol-rich extract from Houttuynia cordata tea. *J Agric Food Chem* 2012, 60:4641-4648

46. Londonkar RL, Poddar PV: Studies on activity of various extracts of Mentha arvensis Linn against drug induced gastric ulcer in mammals. *World J Gastrointest Oncol* 2009, 15:82-88.

47. Gargouri B, Mansour RB, Abdallah FB, Elfekih A, Lassoued S, Khaled H: Protective effect of quercetin against oxidative stress caused by dimethoate in human peripheral blood lymphocytes. *Lipids Health Dis* 2011, 10:149-152

Antioxidant, Cell-protective, and Anti-melanogenic Activities of Leaf Extracts from Wild Bitter Melon (Momordica charantia Linn. var.abbreviata Ser.) Cultivars

Tsung-Hsien Tsai[1], Ching-Jang Huang[2], Wen-Huey Wu[3], Wen-Cheng Huang[3], Jong-Ho Chyuan[4], and Po-Jung Tsai[3]

[1]Department of Dermatology, Taipei Municipal Wan Fang Hospital and Taipei Medical University, Taipei, Taiwan

[2]Institute of Microbiology and Biochemistry, and Department of Biochemical Science and Technology, National Taiwan University, Taipei, Taiwan

[3]Department of Human Development and Family Studies, National Taiwan Normal University, 162 Hoping E. Rd., Sec. 1, Taipei 10610, Taiwan

[4]Hualien District Agricultural Research and Extension Station, Hualien, Taiwan

ABSTRACT

Background

Several wild bitter melon (WBM; *Momordica charantia* Linn. var. *abbreviata* Ser.) cultivars were developed in Taiwan. However, little information is available regarding biological function of WBM leaf. Therefore, the objectives of this study were to investigate the nutrient content, antioxidant, cell protection and anti-melanogenic properties of wild bitter melon leaf.

Results

Methanolic leaf extracts were prepared from a variety and two cultivars of WBM. All extracts exerted potent nitric oxide and hydroxyl radical scavenging capacities. Furthermore, all extracts effectively reduce the production of reactive oxygen species and prevent cell death in UVB-irradiated HaCaT keratinocytes. The cell protective effect of leaf extract was also investigated by the prevention of HaCaT cells from sodium nitroprusside or menadione-induced toxicity, and significant cyto-protective activities were observed for all of them. Additionally, all extracts significantly suppressed tyrosinase activity and melanin levels in B16-F10 melanocytes.

Conclusions

WBM leaf extract showed significant antioxidant, cyto-protective and anti-melanogenic activities. These findings suggested that WBM leaves may be beneficial for preventing the photo-oxidative damage and melanogenesis of skin.

BACKGROUND

Oxidative stress has been thought to play an important role in the pathogenesis of diseases such as cancer, cardiovascular disease, atherosclerosis, diabetes mellitus, and neurodegenerative disorders (Valko et al. [2007]). Ultraviolet (UV) irradiation is the most well-known environmental skin aggressor. Skin, the largest organ of human body, is a physiological barrier that protects the organism against pathogens and chemical or physical damage. Exposure of UV leads to increased reactive oxygen species (ROS) production, which alters gene and protein structure and function (Masaki [2010]). ROS include free radicals such as superoxide anion ($O2^{\bullet-}$), hydroxyl radical ($^{\bullet}OH$), and non-radical molecules like hydrogen peroxide (H_2O_2), singlet oxygen ($^1O_2$), nitric oxide (NO), etc. ROS are involved in the pathogenesis of several skin disorders including photosensitivity diseases and some types of cutaneous malignancy (Bickers and Athar [2006]). Additionally, ROS may accelerate aging process and cause uneven pigmentation. UVB radiation has been shown to augment nitric oxide and peroxynitrite formation in keratinocytes (Deliconstmtinos et al. [1996]). NO production may contribute to the regulation of UV-induced pigmentation. NO derived from keratinocytes increases the amount of the melanogenic factor tyrosinase and then induces melanogenesis (Masaki [2010]). Antioxidant agents may therefore play a protective role during the development of ROS-mediated skin disorders.

Wild bitter melon (WBM; *Momordica charantia* L.var. *abbreviata* Seringe) is a variety of bitter melon (*M. charantia*) in Taiwan. WBM fruit is commonly consumed as vegetable and possesses potent antioxidant and free radical scavenging activities (Wu and Ng [2008]). The young shoots and leaves of WBM are traditionally eaten as greens by the Amis, one of the indigenous peoples of Taiwan. The young tender leaves of bitter melon are also eaten as a vegetable in the Philippines and Indonesia. Besides being as a vegetable, the pounded leaves of bitter melons are applied to the body for skin diseases and burns in Malaysia and India (Lim [2012]). In fact, leaf extracts of bitter melon have been demonstrated to have broad-spectrum antimicrobial (Khan and Omoloso [1998]) and potent antioxidant activities (Kubola and Siriamornpun [2008]).

To date, the genetic improvement of WBM has been achieved to improve their agronomic characteristics such as disease resistance, environmental tolerance and fruit quality. Over the years, several WBM cultivars were developed in Taiwan (Lu et al. [2011], [2012]). WBM fruit extract and its components have been shown to possess numerous pharmacological actions including the activation of peroxisome proliferator-activated receptor, antibacterial, anti-inflammatory, and antioxidant activities (Lu et al. [2011], [2012]; Hsu et al. [2012]). But scientific literatures concerning chemical and biological properties of WBM leaves remain limited.

In this study, we intended to analyze the nutrient contents of fresh leaves from a variety and two cultivars of WBM, and to investigate the total phenolic contents and the antioxidant properties of WBM leaf extracts using cell-free and cell-based assays. In addition, the effects of WBM leaf extracts on tyrosinase activity and melanin levels in B16-F10 melanoma cells were determined. Such a study would contribute to the current knowledge relating to the nutrients and biological functions of WBM leaf.

METHODS

Preparation of Extracts

A variety of WBM (WV) and two cultivars of WBM, Hualien-1 (HL-1) and Hualien-2 (HL-2), used in the present study were cultured in the Hualien District Agricultural Research and Extension Station, Hualien, Taiwan. A voucher specimen has been deposited in the Department of Human Development and Family Studies, National Taiwan Normal University. After washing with water, leaves of WBM were air-dried. They were ground by a blender and then extracted with methanol. Briefly, 10 g of fine-ground WBM leaves was extracted with 100 mL of methanol at room temperature for 4 h. After extraction, the mixture was filtered, and the residue was re-extracted with 100 mL of fresh methanol by stirring overnight. The combined methanol solutions were centrifuged at $12,000 \times g$ for 10 min and evaporated on a rotary evaporator to get methanolic extracts. The methanolic extracts were reconstituted in dimethyl sulfoxide (DMSO) to a concentration of 400 mg/mL for the subsequent experiments. The yields of WV, HL-1, and HL-2 extracts were 29.1%, 22.3%, and 23.5%, respectively.

Nutrient Content Determination of Fresh WBM Leaves

Moisture, crude fat, crude protein, crude fiber and ash determinations of WBM fresh leaves were conducted following the procedure of the AOAC "Official Methods of Analysis" 14th Ed et al. ([1984]). Vitamin C was determined by a colorimetric method of Zhang et al. ([2009]), with modifications. Two-gram dried ground WBM leaf was stirred with 30 mL distilled water for 30 min at room temperature. The homogenization was filtered and then centrifuged at $3000 \times g$ for 15 min. The supernatant was collected and diluted with distilled water to total of 50 mL. A 10:1 dilution was made by taking 1.0 mL of this solution and adding distilled water to 10 mL. A 2 mL volume of trichloro-acetic acid (10%) was added to this suspension and placed for 5 min in an ice bath. Afterwards, 2 mL of Folin–Ciocalteu's phenol reagent was added and vortexed. After 10 min at room temperature, the absorbance was measured at 760 nm against distilled water as a blank. The vitamin C content was estimated through the calibration curve of ascorbic acid.

Total Phenolic Content of WBM Leaf Extract

Total phenolic content of leaf extracts was evaluated using spectrophotometric analysis with Folin-Ciocalteu reagent as described by Tsai et al. ([2008]). Briefly, Folin–Ciocalteu phenol reagent was added to the reconstituted samples and held for 3 min. Then 2 ml of 10% (w/v) sodium carbonate solution were added and allowed to stand at room temperature for 30 min. The absorbance at 765 nm was measured. The total phenolic content was calculated by a standard curve prepared with gallic acid and expressed as milligrams of gallic acid equivalents (GAE) per gram of solid of extract.

Determination of DPPH Radical-Scavenging Activity of WBM Leaf Extracts

The 2, 2-diphenyl-1-picrylhydrazyl (DPPH•) radical-scavenging capacity of each extract was measured as described earlier (Tsai et al. [2008]). Briefly, 20 µL of each sample or 100% DMSO (as a negative control) were allowed to react with 200 µL of freshly prepared 200 µM DPPH ethanolic solution in a 96-well microplate. The reaction mixture was mixed and left to stand for 10 min. The absorbance at 540 nm was determined against a blank of DMSO. The DPPH radical-scavenging activity of WBM extract was calculated as follows: $(1-[A_{sample} - A_{blank\ of\ sample}/A_{DMSO} - A_{blank\ of\ DMSO}]) \times 100\%$.

Determination of No-Scavenging Activity of WBM Leaf Extracts

The NO-scavenging activity of each extract was also measured. Nitric oxide was generated from sodium nitroprusside (SNP) and measured by the Griess reagent (Sumanont et al. [2004]). Briefly, 50 µL of serial diluted sample extract (0.5 ~ 20 mg/mL) was pipetted into a 96-well flat-bottomed plate. Following this, 50 µL of 10 mM sodium nitroprusside dissolved in PBS was added into each well and the plate was then incubated under light at room temperature for 150 min. Finally, an equal volume of Griess reagent was added into each well to measure the nitrite content. The NO-scavenging activity of WBM leaf extract was calculated as follows: $(A_{DMSO} - A_{sample})/A_{DMSO} \times 100\%$.

Determination of Superoxide-Scavenging Activity of WBM Leaf Extracts

The superoxide scavenging activity of WBM leaf extract was measured using non-enzymatic generation of superoxide anions (Robak and Gryglewski [1988]). Briefly, the reaction mixture contained various concentrations of leaf extracts (0.5-20 mg/mL), 80 µM phenazine methosulphate, 624 µM NADH and 200 µM nitro blue tetrazolium (NBT) in phosphate buffer after 15 min of the incubation at room temperature. The result was measured at 560 nm against blank samples without NADH. The percentage of scavenging effect was expressed as % of scavenging activity $= 1 - [A_{560}$ sample $-A_{560}$ blank of sample/A_{560} control $-A_{560}$ blank of control] $\times 100\%$.

Determination of Hydroxyl Radical-Scavenging Activity of WBM Leaf Extracts

The scavenging activity of leaf extracts on hydroxyl radicals produced by the Fenton reaction was evaluated with their quenching effects on the chemical luminescence (CL) signal of the $Fe(II)-H_2O_2$–luminol system (Cheng et al. [2003]). Reaction mixtures (200 µL) included luminol (4 µM), Fe^{2+} (4.6 µM)-EDTA (2.3 µM), H_2O_2 (24 mM), and tested samples (0.02-1 mg/mL of WBM leaf extract). The reaction was initiated by adding Fe^{2+}-EDTA, luminol followed by H_2O_2. The chemiluminescent reaction was performed in a KH_2PO_4-NaOH buffer (pH 7.5) at room temperature. Luminescence intensity was monitored over wavelengths at 460 nm with Synergy HT multidetection microplate reader (Biotek Instruments, Winooski, VT, USA). The CL peak values were recorded in the absence (I_0) or presence (Ii) of leaf extracts. The inhibitory rate (IR) was calculated as $IR = (1 - I_i / I_0) \times 100\%$.

UVB-Irradiated HaCaT Keratinocytes

The immortalized human keratinocyte cell line HaCaT was maintained in Dulbecco's modified Eagle's medium (DMEM, Gibco, Carlsbad, CA, USA) supplemented with 10% heated-inactivated fetal bovine serum (FBS), penicillin (100 U/mL), and streptomycin (100 µg/mL). These cells were incubated at 37°C in a humidified atmosphere with 5% CO_2. The probe 2', 7'-dichlorofluorescein diacetate (H_2DCF-DA; Sigma, St. Louis, MO, USA) was used to monitor the intracellular ROS generation. HaCaT cells (2×10^5 cells/well) were seeded on 6-well plates for 24 h incubation and then pre-treated with WBM leaf extracts at the indicated concentrations for additional 24-h incubation. After incubation, cells were washed with PBS and then irradiated with 80 mJ/cm^2 of UVB (Vilber Lourmat, France). In parallel, non-irradiated cells were treated similarly and were kept in the dark in an incubator for the time of UVB treatment. After UVB irradiation, cells were harvested and washed twice with PBS, and then re-suspended in 10 µM H_2DCF-DA at

37°C for 30 min incubation. Stained cells were washed with PBS and re-suspended in PBS. ROS generation of HaCaT cells was determined by flow cytometry (FACscan, Hercules, CA, USA) using 488 nm for excitation and 525 nm for emission. Mean fluorescence intensity (MFI) detected by FLl channel was analyzed using the Windows Multiple Document Interface software (WinMDI 2.8). Data were expressed as percentages of vehicle control values.

To determine the protective effect of WBM leaf extracts against UVB-induced cytotoxicity, HaCaT cells were seeded in 96-well plates at a density of 2 × 10^4 cells/well for 24 h incubation and then pre-treated with WBM leaf extracts at the indicated concentrations for additional 24-h incubation. After incubation, cells were washed with PBS and then irradiated with 80 mJ/cm^2 of UVB. In parallel, non-irradiated cells were treated similarly and were kept in the dark in an incubator for the time of UVB treatment. After challenge of HaCaT cells with UVB, the cells were incubated in fresh DMEM with 10% FBS at 37°C for 30 min or 16 h, and collected for further analysis. The 3-(4, 5-dimethylthiazol-2-yl)-2, 5-diphenyltetrazolium bromide (MTT) assay was performed to determine cellular viability.

Cell Protection Effect of WBM Leaf Extracts Against SNP-induced Toxicity

The protective effect of WBM leaf extract against cell death induced by sodium nitroprusside (SNP) was determined by the method of Bastianetto et al. ([2010]). SNP is a well-known NO releasing with purported toxic and apoptotic effects in keratinocytes (Bastianetto et al. [2010]). SNP-induced toxicity was performed in HaCaT cells (5 × 10^4/well) plated in 96 wells. After 24 h, the medium was removed and replaced with medium contain SNP (2 mM) in the presence or absence of leaf extracts (50 ~ 200 μg/mL). Cell viability was determined 24 h later using the MTT assay.

Cell Protection Effect of WBM Leaf Extracts Against Menadione-Induced Toxicity

The protective effect of WBM leaf extract against cell death induced by menadione (2-methyl-1, 4-naphthoquinone) was measured by the method of Klausc et al. ([2010]). Menadione is highly cytotoxic, strongly induced ROS formation in human HaCaT keratinocytes. Menadione-induced toxicity was performed in HaCaT cells (5×10^4/well) plated in 96 wells. After 24-h incubation, the medium was removed and replaced with medium contain menadione (50 µM) in the presence or absence of WBM leaf extracts (50~200 µg/mL). Cell viability was determined 24 h later using the MTT assays.

Determination of Cellular Tyrosinase Activity in Melanoma Cells

The established murine B16-F10 melanoma cell line offers a melanogenesis model (Yokozawa and Kim [2007]). The B16-F10 cell line (BCRC 60031) was obtained from the Bioresource Collection and Research Center (Hsinchu, Taiwan) and was cultured in DMEM (Gibco) supplemented with 10% heat-inactivated FBS, penicillin (100 U/mL), and streptomycin (100 µg/mL) at 37°C in a humidified atmosphere with 5% CO_2. The effect of WBM leaf extracts on cell viability of B16-F10 melanoma cells was first investigated. B16-F10 melanoma cells (1×10^4 cells/well) were seeded in 96-well culture plates for 24 h prior to use. The cells were treated with various concentrations of 50~250 µg/mL of WBM leaf extracts for 24 h. The cell viability was evaluated using the MTT assays.

The tyrosinase activity in B16-F10 cells was examined by measuring the rate of oxidation of L-DOPA (Yokozawa and Kim [2007]). B16-F10 cells (2.5×10^4 cells/well) were plated in 24-well dishes for 24 h incubation before the use. The cells were then incubated in the presence or absence of 25 ng/mL α-melanocyte stimulating hormone (α-MSH) and various concentrations of WBM leaf extracts for 48 h. In addition, kojic acid [5-hydroxy-2-(hydroxymethyl)-4-pyrone], a popular inhibitor of tyrosinase (Kahn [1995]) was used as a positive control. The cells were lysed in 200 µL of 50 mM sodium phosphate buffer (pH 6.8) containing

1% Triton X-100 and 0.1 mM phenylmethylsulfonyl fluoride and then frozen at −80°C for 30 min. After thawing and mixing, cellular extracts were clarified by centrifugation at 12,000×g for 30 min at 4°C. The supernatant (80 μL) and 20 μL of L-DOPA (2 mg/mL) were placed in a 96-well plate, and the absorbance at 492 nm was read for 30 min at 37°C using a micro-plate reader.

Determination of Melanin Content in Melanoma Cells

Melanin content was measured as described previously (Aoki et al. [2007]) with modifications. B16-F10 cells were incubated in the presence or absence of 25 ng/mL α-MSH and various concentrations (50~200 μg/mL) of leaf extracts for 48 h. The cells were rendered soluble in 1 N NaOH containing 10% DMSO at 60°C for 30 min, and then 200 μL portions of crude cell lysate was transferred into 96-well plates. Melanin concentration was calculated by the absorbance measured at 405 nm through the calibration curve of melanin.

Statistical Analysis

All data are presented as the mean ± standard deviation (SD). Statistical analyses were performed using Statistical Package of Social Science version 17.0 for Windows (SPSS Inc., Chicago, Illinois, USA). Analysis of variance was performed by ANOVA procedures. Significant differences between means between groups were determined using Duncan's multiple range tests at a level of $p < 0.05$.

RESULTS

Nutrient Content of WBM Leaves

Moisture, crude protein, crude fat, crude fiber, and ash content (%) of WBM leaves were reported on dry-weight basis and given in Table 1. In fresh WBM leaves, crude fat, crude protein, and crude fiber contents ranged from 0.74% to 1.22%, 4.22% to 6.26%, and 1.59% to 1.92%

on dry-weight basis, respectively. Ash contents in WBM leaves ranged from 2.5% to 3.76% on dry-weight basis. Vitamin C contents in WBM leaves ranged from 1647.3 to 2059.0 µg/g dry basis.

Table 1: Mositure, crude protein, crude fat, crude fiber, and ash content and vitamin C of fresh wild bitter melon leaves (on dry-weight basis)

Variety and cultivars	Moisture (%)	Crude protein (%)	Crude fat (%)	Crude fiber (%)	Ash (%)	Vitamin C (µg/g dry basis)
WV	83.20	5.28	0.99	1.90	3.76	1647.32
HL-1	81.22	6.26	1.22	1.92	2.85	1920.14
HL-2	83.77	4.22	0.74	1.59	2.50	2059.03

Tsai et al.

Tsai *et al Botanical Studies* 2014 **55**:78, doi: 10.1186/s40529-014-0078-y

Total Phenolic Content of WBM Leaf Extract

The total phenolic content of WBM leaf extracts was determined as gallic acid equivalents (Table 2). Among three extracts tested, WV had the highest phenolic content (34 mg GAE/g extract), followed by HL-1 and HL-2 (26 mg GAE/g extract).

Table 2: Total phenolics content and radical-scavenging capacity of methanolic extracts from wild bitter melon leaves

Variety and cultivars	Total phenolics (mg GAE/g)	IC_{50} values (mg/mL)			
		DPPH	NO	Superoxide	Hydroxyl radical
WV	33.70±0.48[b]	4.79±0.30[a]	0.79±0.01	5.77±0.10[a]	0.026±0.003[a]
HL-1	25.86±0.36[a]	28.00±0.83[c]	0.76±0.02	9.12±0.22[c]	0.042±0.004[b]
HL-2	25.94±0.35[a]	20.39±1.12[b]	0.80±0.02	7.71±0.35[b]	0.022±0.004[a]

GAE: Gallic acid equivalent. Data are expressed as the mean ± SD. Values in a column followed by the same superscript letter are not significantly different as determined by Duncan's multiple tests.

Tsai *et al. Botanical Studies* 2014 55:78, doi: 10.1186/s40529-014-0078-y

Free Radical-Scavenging Activity of WBM Extracts

The free radical-scavenging properties of leaf extracts were examined (Figure 1) and then expressed as the IC_{50} which is the concentration of leaf extract that causes 50% inhibition of respective radical generation (Table 2). The DPPH radical scavenging assay is a simple and widely used screening for bioactive compound discovery. The scavenging activity of WBM leaf extracts on DPPH radicals is in a concentration-dependent manner (Figure 1A). The IC_{50} values of WV, HL-1, and HL-2 extracts on DPPH scavenging activity were 4.79, 28.0, and 20.39 mg/mL, respectively (Table 2). Among tested leaf extracts, WV is the strongest DPPH radicals' scavenger. HL-1 extract is the least potent scavenger of DPPH.

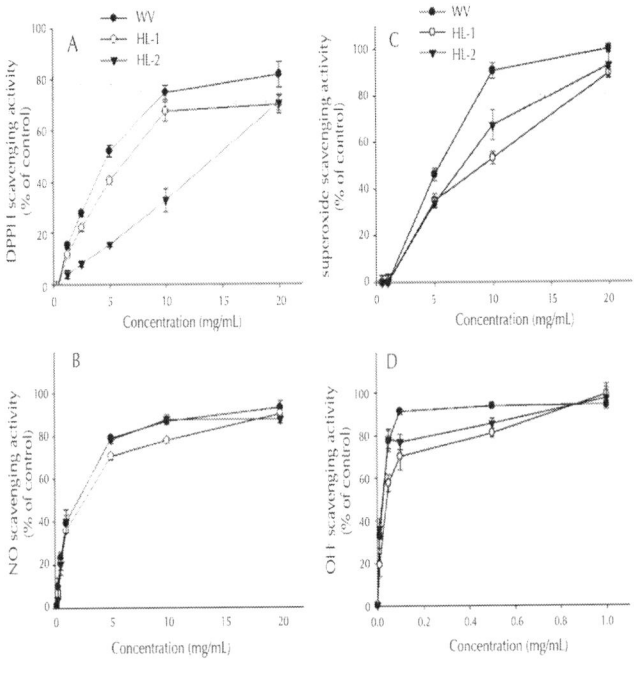

Figure 1: Scavenging activities of WBM leaf extracts against the DPPH (A), nitric oxide (B), superoxide anion (C), and hydroxyl radical (D). Data are given as the mean ± SD (n = 3). Radical-scavenging capacity of leaf extract was represented as % of vehicle control.

The NO-scavenging activity of leaf extracts were measured to have IC_{50} values of 0.79, 0.76, and 0.80 mg/mL for WV, HL-1, and HL-2 extracts, respectively (Figure 1B, Table 2), where there was no statistically significant difference observed among the variety and cultivars.

WBM leaf extracts also caused a concentration-dependent inhibition of superoxide anion (Figure 1C). The respective IC_{50} values of WV, HL-1, and HL-2 extracts on superoxide scavenging activity were 5.77, 9.12, and 7.71 mg/mL (Table 2), indicating that WV extract is more potent scavenger of superoxide anion than others.

The hydroxyl radical-scavenging capacity of leaf extracts is shown in Figure 1D. The respective IC_{50} values of WV, HL-1, and HL-2 extracts were 26, 42, and 22 µg/mL (Table 2). WV and HL-2 had more effective scavenging activity than HL-1.

WBM Leaf Extracts Inhibit UV-induced ROS Production and Keratinocyte Death

As shown in Figure 2A, UVB exposure of HaCaT keratinocytes resulted in an immediate and significant elevation of intracellular ROS. Treatment of WBM leaf extracts (50 and 100 µg/mL) prior to UVB irradiation significantly inhibited intracellular ROS generation. All leaf extracts effectively inhibited UVB-induced ROS production. After UVB exposure, HaCaT cells were further incubated at 37°C for 30 min (Figure 2B) or 16 h (Figure 2C). The results showed that UVB irradiation did not significantly affect cell viability for further 30 min incubation. All three WBM leaf extracts did not significantly modulate cell viability of HaCaT cells during 30 min incubation (Figure 2B). WBM leaf extract-pretreated HaCaT cells were exposed to UVB and incubated for an additional 16 h. MTT assay showed that cell viability of HaCaT cells was significantly decreased after UVB exposure (Figure 2C). Notably, all three WBM leaf extracts at concentrations of 25, 50, and 100 µg/mL revealed a protective effect on the viability of irradiated HaCaT cells (Figure 2C).

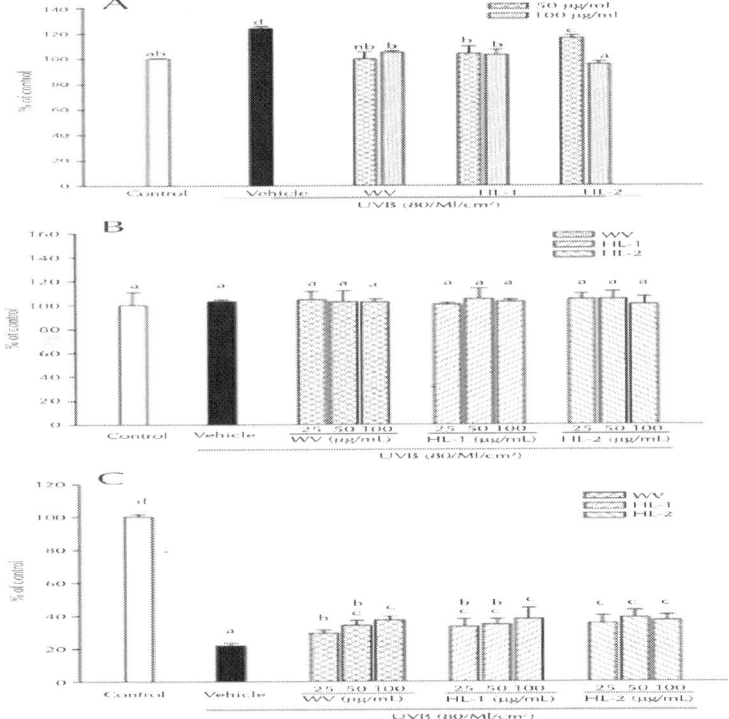

Figure 2: Effects of WBM leaf extracts on UV-induced ROS production and cytotoxicity in HaCaT keratinocytes. After exposure of UVB (80 mJ/cm²), the cells were further incubated at 37°C for 30 min and then ROS generation was determined by flow cytometry. ROS production was represented as % of control without UVB irradiation. (A) After exposure of UVB (80 mJ/cm²), the cells were further incubated at 37°C for 30 min (B) or 16 h (C) and then cell viability was measured using the MTT assay. Non-irradiated HaCaT keratinocytes were used as the control. Cell viability is expressed as the percentage of control. Data are presented as the mean ± SD of triplicate determinations. Values with the same letter are not significantly different as determined by Duncan's multiple range tests.

Cellular Protection Effect of WBM Leaf Extracts against Oxidants

Prior to the determination of cellular protection effect, the cytotoxic effect of WBM leaf extracts on HaCaT keratinocytes were examined.

HaCaT cells were incubated with WBM leaf extracts at various concentrations for 24-h incubation. The cell viability was determined by MTT methods. All three WBM leaf extracts (up to 300 μg/mL) had no cytotoxic effect on HaCaT cells (data not shown). On the other hand, treatment of HaCaT cells with SNP (2 mM) (Figure 3A) and menadione (50 μM) (Figure 3B) strongly impaired cell viability. To determine the cyto-protection activities, HaCaT keratinocytes were simultaneously exposed to cytotoxic agents either SNP or menadione as well as WBM leaf extracts at various concentrations (Figure 3). All three WBM leaf extracts strongly attenuated SNP-induced toxicity at a concentration of 50 μg/mL (Figure 3A). The HL-2 extract was the most effective one in protecting cells against SNP-induced toxicity with an EC_{50} (effective concentrations) of 31.31 μg/mL, followed by WV (EC_{50} = 49.35 μg/mL), and HL-1 (EC_{50} = 84.72 μg/mL).

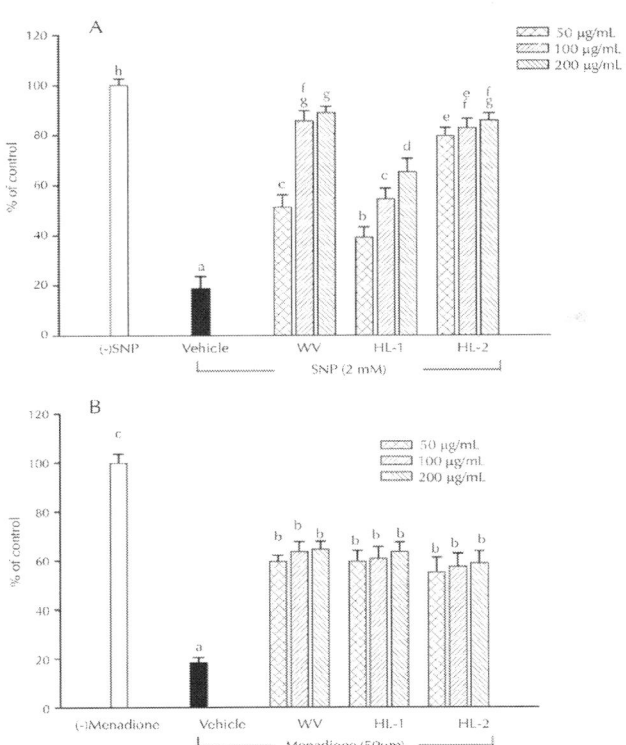

Figure 3: Effects of WBM leaf extracts against keratinocyte death induced by SNP (A) and menadione (B) treatment. Cells were exposed to either SNP

(2 mM) or menadione (50 μM) in the absence (vehicle) or presence of leaf extracts (50–200 μg/mL). Cell viability was determined 24 hours later using MTT assay. Data are presented as the mean ± SD of triplicate determinations. Values with the same letter are not significantly different asdetermined by Duncan's multiple range tests.

We then performed a cell viability assay to evaluate the capacity of WBM leaf extracts to protect HaCaT cells against the toxicity induced by the ROS releasing menadione (Figure 3B). Exposure of HaCaT cells to menadione resulted in cell death which was reduced by WV (EC_{50} = 42.05 μg/mL), HL-1 (EC_{50} = 42.07 μg/mL), and HL-2 (EC_{50} = 43.55 μg/mL). According to EC_{50} values, the three varieties were comparable in their cyto-protective effect against menadione-induced oxidative stress.

Anti-tyrosinase and Anti-melanogenic Properties of WBM Leaf Extracts

To investigate the anti-melanogenic activity in cellular system, the cytotoxicity of WBM leaf extracts on B16-F10 cells was first evaluated after incubated with extracts for 24 hr. Our data showed that these three extracts (up to 200 μg/mL) did not show any cytotoxic effect on B16-F10 cells (data not shown). Thus, WBM leaf extract (50–200 μg/mL) was used to examine their anti-tyrosinase activity and intracellular melanin formation. As shown in Figure 4, all three WBM leaf extracts exhibited significantly inhibitory effect on tyrosinase activity (Figure 4A). However, at a concentration of 50 μg/mL, the anti-tyrosinase effect of WV was not significant. At a concentration of 200 μg/mL, HL-1 extract exhibited more significant anti-tyrosinase activity than others. The inhibitory effect of HL-1 extract (ranged from 50 to 200 μg/mL) on tyrosinase activity were similar to kojic acid, a well-known tyrosinase inhibitor (Figure 4A).

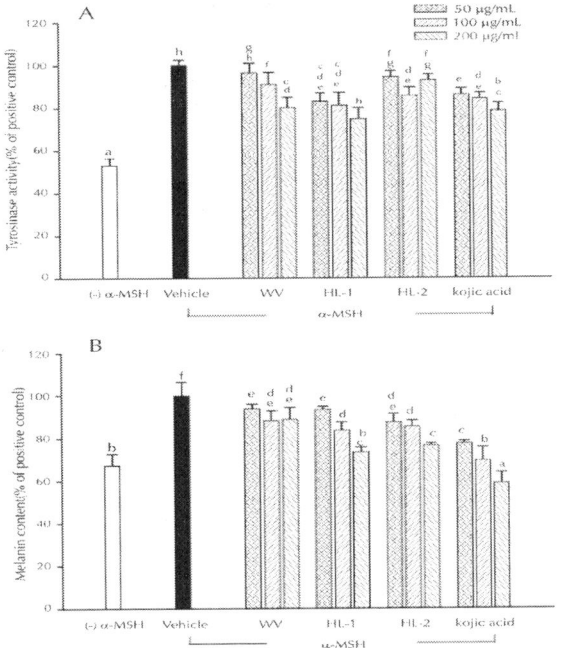

Figure 4: Effects of WBM leaf extract on cellular tyrosinase activity (A) and melanin level (B) of α-MSH treated B16-F10 melanocytes. Each value is expressed as the mean ± SD (n = 3). Values with the same letter are not significantly different as determined by Duncan's multiple range tests.

All three WBM leaf extracts significantly suppressed melanin formation in B16-F10 cells (Figure 4B). At a concentration of 200 µg/mL, WV, HL-1, and HL-2 extracts and kojic acid reduced melanin production of 11.29%, 26.63%, 23.44% and 41.2%, respectively. The anti-melanogenis effect of HL-1 is higher than those of WV and HL-2, but lower than that of kojic acid.

DISCUSSION

In this study, the cultivars of wild bitter melon leaves are proved in possessing antioxidant and anti-melanogenic properties. Although there are some differences in the degree they exerted, however, all of them possess antioxidant to reduce UVB-induced ROS generation and prevent cellular death against oxidants in keratinocytes.

Leafy vegetables are good sources of not only minerals but also vitamins, antioxidants and pigments. The tender leaf tips and leaves of bitter melon are a rich source of minerals, vitamin C, folic acid and vitamin A (Zhang et al. [2009]; Lim [2012]). This study analyzed nutrient content of WBM leaf and provided further information on nutrient contents on WBM leaf for nutritionists and the general public. During the past few decades, the uses of natural antioxidants and plant extracts for human health have received increasing attention.

The findings presented here showed that WBM leaf extracts exhibited antioxidant properties because of their capacity to scavenge various free radicals and to reduce oxidant-induced cellular death. As Figure 1 and Table 2 shown, WBM leaf extracts exerted hydroxyl radical and NO scavenging activities. The most impressive radical-scavenging activity is against hydroxyl radical, which is considered to be the most reactive one among ROS. The IC_{50} values for hydroxyl radial ranged from 22 to 42 µg/mL. The wild variety and HL-2 were superior to HL-1. Leaf extract of Thai bitter melon possesses hydroxyl-radical scavenging activity with an IC_{50} value of 167 ± 0.96 mg/mL (Kubola and Siriamornpun [2008]). Lu et al. ([2012]) demonstrated fruit extract of the most effective WBM cultivar in Taiwan exerted potent hydroxyl-radical scavenging activity (IC_{50}, 37 µg/mL). These results further supported that leaf and fruit of WBM possess strong hydroxyl-radical scavenging activity.

Numerous phytochemicals such as momordicine, kuguacins, and phenolics, have been isolated from bitter melon leaves (Lim [2012]; Kubola and Siriamornpun [2008]). The new finding triterpenoids isolated from the stems and fruits of bitter melons show their antioxidant property (Liu et al. [2010]; Lin et al. [2011]). Kubola and Siriamornpun ([2008]) reported that leaf extract of bitter melon possess antioxidant activity, based on DPPH radical-scavenging activity and ferric acid reducing power. The predominant phenolic compounds in the leaf of bitter melon are gallic acid, followed by caffeic acid and catechin (Kubola and Siriamornpun [2008]). We recently investigated phenolic compounds of these three WBM leaf extracts using HPLC methods (Kubola and Siriamornpun [2008]; Zhang et al. [2009]) with some modifications. The phenolic compounds found in WBM leaves were gallic acid, salicylic acid, cinnamic acid, myricetin, quercetin, and luteolin Because of the diversity and complexity of the natural mixtures of antioxidants in WBM leaf extracts, it is rather difficult to characterize all of compounds by HPLC. Further work is still required to verify the anti-oxidative constituents of WBM leaves.

Skin is constantly exposing to environmental insults, among all, UV light is thought to be the most harmful one. UV exposure can cause oxidative stress and inflammation. Eating plenty of green leafy vegetables has been considered to be beneficial for photoprotection (Mukhtar [2003]). Phytonutrients with antioxidant activity have been considered to be beneficial for attenuating UV-caused oxidative stress and oxidative stress-mediated skin disorders (Evans and Johnson [2010]). To elucidate the antioxidant activity exerted by WBM leaf extracts, the intracellular ROS generation in UV-irradiated keratinocyte was carried out using an oxidant-sensitive fluorescent probe DCF-DA. By measuring the intercellular ROS scavenging activity of WBM leaf extracts, the results has shown that pre-treatment of extracts could reduce the intracellular ROS production induced by UV within cells. Moreover, all three WBM leaf extracts showed the protective effect on the viability of UVB-irradiated keratinocytes (Figure 2). Therefore, WBM leaf extract could be proposed as a photo-protective agent preventing harmful effects of UVB exposure on human keratinocytes.

As WBM leaf extracts showed antioxidant effects against free radicals and intercellular ROS, their cell protective effects on keratinocytes against the cellular damage induced by SNP (a NO donor) and menadione (a ROS donor) were evaluated (Figure 3). The cell protective property of WBM leaf extracts against SNP or menadione-induced cell death can be observed herein. Kumar et al. ([2010]) reported that fruit extract of bitter melon showed significant cytoprotection against oxidants on fibroblasts and keratinocytes. In this study, WBM leaf extract protect HaCaT cells from damage caused by SNP, which implies that WBM leaf extracts are able to block the harmful events induced by NO overproduction, a process relevant to premature skin aging occurring upon long term UV exposure (Weller [2003]). Similarity, all three WBM leaf extracts possessed cyto-protection against menadione-induced damage on HaCaT keratinocytes, suggesting its ability to reduce superoxide-mediated stress in skin. In considering that ROS plays an important role in the pathogenesis of many skin disorders and chronologic skin aging (Bickers and Athar [2006]), these findings presented in this study suggest that WBM leaf is a good source of natural antioxidants and may prevent the ROS-mediated skin disorders.

Clinical changes recognizable as photoaging include wrinkle, inelasticity, telangiectasia and pigmentary change. Although not detrimental in nature, hyperpigmentation can be a big cosmetic concern, especially among Asian. The inhibition of melanogenesis is a critical target for skin-whitening cosmetic and treatment of abnormal pigmentation (Yokozawa and Kim [2007]). Since tyrosinase is a rate-limiting enzyme of melanogenesis, inhibition on its action is one major strategy to treat hyperpigmentation. The anti-tyrosinase and anti-melanogenic activity of natural substances have been studied extensively. Fruit extracts of bitter melon have been shown to exert significant inhibitory effect on mushroom tyrosinase (Kamkaen et al. [2007]; Masuda et al. [2007]; Baurin et al. [2002]). Kamkaen et al. ([2007]) reported that methanolic extract of bitter melon fruit can exert a significant mushroom tyrosinase inhibition (78.9% compared to positive control of kojic acid). Masuda et al. ([2007]) reported that ethanol extract of bitter melon fruit can show 21.7% of mushroom tyrosinase inhibitory activity at a concentration of 0.5 mg/mL. Similar finding was reported in previous study that used propylene glycol/deionized water extract of bitter melon plants with 32% inhibition of mushroom tyrosinase (Baurin et al. [2002]). In our preliminary study, WBM leaf extracts exhibited suppressive effect on mushroom tyrosinase activity at a concentration of 5 mg/mL. It is not yet known whether WBM leaf extract has anti-melanogenic activity in cellular assay. In the experiments reported here, HL-1 showed more potent anti-tyrosinase and anti-melanogenic activities than others. Regarding the inhibitory effect on tyrosinase activity, HL-1 extract was similar to kojic acid at each tested concentration (Figure 4). Kojic acid shows its anti-tyrosinase activity by chelating the copper ion in the active site of tyrosinase (Kahn [1995]). However, the inhibitory mechanism or effective constituents of WBM leaf extract is still unclear. Kim et al. ([2008]), Yokozawa and Kim ([2007]), and Kim ([2007]) reported that the antimelanogenic action of some agents is related to their antioxidant activity. Concerning the action mechanism of anti-tyrosinase agents, blocking of oxidative pathway (Kim [2007]) and binding of enzyme activity sites (Curto et al. [1999]) may be involved in inhibiting the catalytic reaction of tyrosinase. Previous studies demonstrated that gallate and its derivatives (Kim [2007]; Tanford [1980]) and p-alkoxybenzoic acid derivatives (Chen et al. [2005]) can act as tyrosinase inhibitors. The bulky hydrophobic portions of gallate derivatives were considered to interact with tyrosinase's hydrophobic protein bulk surrounding the binuclear copper active site and thus gallate derivatives exerted the anti-tyrosinase effect (Tanford [1980]). Since gallic acid was found in WBM leaf extracts, it speculated that gallic acid may contribute, at least partially, to anti-melanogenic activity of WBM leaf extracts. Other unknown active constituents present in WBM leaf extract could also play critical roles in the biological effects. The identification of these functional or bioactive ingredients in WBM leaf extract is an interesting topic for future study.

CONCLUSIONS

In summary, all leaf extracts form WBM variety and cultivars showed antioxidant, cell protection, and anti-melanogenic activities. Among them, HL-1 showed the most potent anti-melanogenic activity. These findings may be used to develop health foods or cosmetics, and to increase the range of applications of traditional agricultural vegetables.

ACKNOWLEDGEMENTS

We thank the excellent technical assistance of Technology Commons, College of Life Science, and National Taiwan University, Taiwan with flow cytometry. This work was supported by research grants (NSC 98-2321-B-003-001 and NSC 99-2321-B-003-001) from the National Science Council, Taipei, Taiwan.

REFERENCES

1.   AOAC "Official Methods of Analysis" 14th ed., Washington, D.C.: Association of Official Analytical Chemists; 1984.

2.   Aoki Y, Tanigawa T, Abe H, Fujiwara Y (2007) Melanogenesis inhibition by an oolong tea extract in B16 mouse melanoma cells and UV-induced skin pigmentation in brownish guinea pigs. Biosci Biotechnol Biochem 71:1879-1885

3.   Bastianetto S, Dumont Y, Duranton A, Vercauteren F, Breton L, Quirion R (2010) Protective action of resveratrol in human skin:

possible involvement of specific receptor binding sites. PLoS One 5(9):e12935

4.   Baurin N, Arnoult E, Scior T, Do QT, Bernard P (2002) Preliminary screening of some tropical plants for anti-tyrosinase activity. J Ethnopharmacol 82:155-158

5.   Bickers DR, Athar M (2006) Oxidative stress in the pathogenesis of skin disease. J Investig Dermatol 126:2565-2575

6.   Chen QX, Song KK, Qiu L, Liu XD, Huang H, Guo HY (2005) Inhibitory effects on mushroom tyrosinase by p-alkoxybenzoic acids. Food Chem 91:269-274

7.   Cheng Z, Yan G, Li Y, Chang W (2003) Determination of antioxidant activity of phenolic antioxidants in a fenton-type reaction system by chemiluminescence assay. Anal Bioanal Chem 375:376-380

8.   Curto EV, Kwong C, Hermersdorfer H, Glatt H, Santis C, Virador V (1999) Inhibitors of mammalian melanocyte tyrosinase: *in vitro* comparisons of alkyl esters of gentisic acid with other putative inhibitors. Biochem Pharmacol 57:633-672

9.   Deliconstmtinos G, Villiotou V, Stawides JC (1996) Alterations of nitric oxide synthase and xanthine oxidase activities of human keratinocytes by ultraviolet B radiation. Biochem Pharmacol 51:1727-1738

10.   Evans JA, Johnson EJ (2010) the role of phytonutrients in skin health. Nutrients 2:903-928

11.   Hsu C, Tsai TH, Li YY, Wu WH, Huang CJ, Tsai PJ (2012) Wild bitter melon *(Momordica charantia* Linn. var. *abbreviata* Ser.) extract and its bioactive components suppress *Propionibacterium acnes*-induced inflammation. Food Chem 135:976-984

12.   Kahn V (1995) Effect of kojic acid on the oxidation of L-DOPA, norepinephrine, and dopamine by mushroom tyrosinase. Pigment Cell Res 8:234-240

13.   Kamkaen N, Mulsri N, Treesak C (2007) Screening of some tropical vegetables for anti-tyrosinase activity. Thail Pharm Health Sci J 2:15-19

14.   Khan MR, Omoloso AD (1998) *Momordica charantia* and *Allium sativum*: broad spectrum antibacterial activity. Korean J Pharmacogn 29:155-158

15. Kim YJ (2007) Anti-melanogenic and antioxidant properties of gallic acid. Biol Pharm Bull 30:1052-1055

16. Kim YJ, Kang KS, Yokozawa T (2008) the anti-melanogenic effect of pycnogenol by its anti-oxidative actions. Food Chem Toxicol 46:2466-2471

17. Klausc V, Hartmannb T, Gambinid J, Grafb P, Stahlb W, Hartwigc A, Klotza LO (2010) 1, 4-Naphthoquinones as inducers of oxidative damage and stress signaling in HaCaT human keratinocytes. Arch Biochem Biophys 496:93-100

18. Kubola J, Siriamornpun S (2008) Phenolic contents and antioxidant activities of bitter gourd (*Momordica charantia* L.) leaf, stem and fruit fraction extracts in vitro. Food Chem 110:881-890

19. Kumar R, Balaji S, Sripriya R, Nithya N, Uma TS, Sehgal PK (2010) In vitro evaluation of antioxidants of fruit extract of *Momordica charantia* L. on fibroblasts and keratinocytes. J Agric Food Chem 58:1518-1522

20. Lim TK (2012) Momordica charantia. In: Lim TK (Ed) Edible medicinal and non-medicinal plants, Springer, New York. pp 331-368 Publisher Full Text

21. Lin KW, Yang SC, Lin CN (2011) Antioxidant constituents from the stems and fruits of *Momordica charantia*. Food Chem 127:609-614

22. Liu CH, Yen MH, Tsang SF, Gan KH, Hsu HY, Lin CN (2010) Antioxidant triterpenoids from the stems of *Momordica charantia*. Food Chem 118:751-756

23. Lu YL, Liu YH, Liang WL, Chyuan JH, Cheng KT, Liang HJ, Hou WC (2011) Antibacterial and cytotoxic activities of different wild bitter gourd cultivars (*Momordica charantia* L. var.*abbreviata* Seringe). Bot Stud 52:427-434

24. Lu YL, Liu TH, Chyuan JH, Cheng KT, Liang WL, Hou WC (2012) Antioxidant activities of different wild bitter gourd (*Momordica charantia* L. var. *abbreviata* Seringe) cultivars. Bot Stud 53:207-214

25. Masaki H (2010) Role of antioxidants in the skin: anti-aging effects. J Dermatol Sci 58:85-90

26. Masuda T, Fujita N, Odaka Y, Takeda Y, Yonemori S, Nakamoto K, Kuninaga H (2007) Tyrosinase inhibitory activity of ethanol

extracts from medicinal and edible plants cultivated in Okinawa and identification of a water-soluble inhibitor from the leaves of *Nandina domestica*. Biosci Biotechnol Biochem 71:2316-2320

27.   Mukhtar H (2003) Eat plenty of green leafy vegetables for photo-protection: emerging evidence. J Invest Dermatol 121:doi:10.1046/j.1523-1747.2003.12377.x.

28.   Robak J, Gryglewski RJ (1988) Flavonoids are scavengers of superoxide anions. Biochem Pharmacol 37:837-84

29.   Sumanont Y, Murakami Y, Tohda M, Vajragupta O, Matsumoto K, Watanabe H (2004) Evaluation of the nitric oxide radical scavenging activity of manganese complexes of curcumin and its derivative. Biol Pharm Bull 27:170-173

30.   Tanford C (1980) the hydrophobic effect: Formation of micelles and biological membranes. John Wiley & Sons, New York.

31.   Tsai TH, Tsai TH, Chien YC, Lee CW, Tsai PJ (2008) *In vitro* antimicrobial activities against cariogenic streptococci and their antioxidant capacities: A comparative study of green tea versus different herbs. Food Chem 110:859-864

32.   Valko M, Leibfritz D, Moncol J, Cronin MT, Mazur M, Telser J (2007) Free radicals and antioxidants in normal physiological functions and human disease. Int J Biochem Cell Biol 39:44-84

33.   Weller R (2003) Nitric oxide: a key mediator in cutaneous physiology. Clin Exp Dermatol 28:511-514

34.   Wu SJ, Ng LT (2008) Antioxidant and free radical scavenging activities of wild bitter melon (*Momordica charantia* Linn. var. *abbreviata* Ser.) in Taiwan. LWT Food Sci Technol 41:323-330

35.   Yokozawa T, Kim YJ (2007) Piceatannol inhibits melanogenesis by its antioxidative actions. Biol Pharm Bull 30:2007-2011

36.   Zhang M, Hettiarachchy NS, Horax R, Chen PY, Kenneth F (2009) Effect of maturity stages and drying methods on the retention of selected nutrients and phytochemicals in bitter melon (*Momordica charantia*) leaf. J Food Sci 74:C441-C448

In Vitro Propagation and Analysis of Secondary Metabolites in Glossogyne Tenuifolia (Hsiang-Ju) - A Medicinal Plant Native to Taiwan

Chia-Chen Chen[1], Hung-Chi Chang[2], Chao-Lin Kuo1,
Dinesh Chandra Agrawal[3], Chi-Rei Wu[1],
and Hsin-Sheng Tsay[3]

[1]Department of Chinese Pharmaceutical Sciences and Chinese Medicine Resources, China Medical University, Taichung, Taiwan

[2]Department of Golden-Ager Industry Management, Chaoyang University of Technology, Taichung, Taiwan

[3]Department of Applied Chemistry, Chaoyang University of Technology, No.168, Gifong E. Rd., Taichung, 41349, Wufong, Taiwan

ABSTRACT

Background

Glossogyne tenuifolia Cassini (Hsiang-Ju in Chinese) is a perennial herb native to Penghu Islands, Taiwan. The herb is a traditional anti-pyretic and hepatoprotective used in Chinese medicine. Several studies on *G. tenuifolia* have demonstrated its pharmacological values of antioxidation, anti-inflammation, immunomodulation, and cytotoxicity on several human cancer cell lines. Active compounds, oleanolic acid and luteolin in *G. tenuifolia* are affected by several factors, including climatic change, pathogens and agricultural practices. Plant population of *G. tenuifolia* has been severely affected and reduced considerably in natural habitat due to the use of herbicides by farmers. Also, collection of plant material from the natural habitat is restricted to a few months in a year. Therefore, the objective of the present study was to develop an efficient micropropagation protocol for *G. tenuifolia*. The study also aimed to investigate the influence of *in vitro* growth environment on the active compounds in *in vitro* shoots, tissue culture raised greenhouse plants; compare the values with wild plants and commercially available crude drug.

Results

Half-strength MS (Murashige and Skoog) basal medium supplemented with 0.1 mg/L 6-benzyladenine (BA) and 0.1 mg/L -naphthaleneacetic acid (NAA) induced the maximum average number of shoots (7.3) per shoot tip explant excised from *in vitro* grown seedlings. Induction of rooting in cent percent *in vitro* shoots with an average number of 6.6 roots/shoot was achieved on ½ strength MS medium supplemented with 3.0 mg/L indole-3-acetic acid (IAA). The rooted plantlets acclimatized successfully in the greenhouse with a 100% survival rate. HPLC analysis revealed that the quantity of oleanolic acid and luteolin in *in vitro* shoots, tissue culture plants in the greenhouse, wild type plants and commercial crude drug varied depending upon the source. The oleanolic acid and luteolin contents were found to be significantly higher (16.89 mg/g and 0.84 mg/g, respectively) in 3-month old tissue culture raised plants in greenhouse compared to commercially available crude drug (6.51 mg/g, 0.13 mg/g, respectively).

Conclusions

We have successfully developed an *in vitro* propagation protocol for *G. tenuifolia* which can expedite its plant production throughout the year. The contents of oleanolic acid and luteolin in the tissue culture raised plants in the greenhouse were significantly higher than the marketed crude drug demonstrating the practical application of the tissue culture technology. These findings may be very useful in micropropagation, germplasm conservation and commercial cultivation of *G. tenuifolia*. So far, there is no published report on tissue culture propagation of this important medicinal plant species.

BACKGROUND

Medicinal herbs have played a significant role in maintaining human health and improving the quality of life for thousands of years. Many active phytochemicals, including flavonoids, terpenoids, lignans, sulfides, polyphenolics, carotenoids, coumarins, saponins, plant sterols, curcumins, and phthalides, have been identified (Craig [1999]). Some of these phytochemicals have been found to be potent antioxidants, metal chelators, or free radical scavengers, which may account for their health promoting properties (Cotell et al. [1996]). Today, medicinal plants are important to the global economy as approximately 85% of traditional medicine preparations involve the use of plants or plant extracts (Vieira and Skorupa [1993]). In the past few decades, there has been a resurging interest in the study and use of medicinal plants in health care and in recognition of the importance of medicinal plants to the health system (Hoareau and DaSilva [1999]). This has led to an exponential rise in demand for herbal medicines, and also a considerable international awareness about the dwindling supply of the world's medicinal plants. Therefore, all possible modes of plant propagation and large scale cultivation have been explored. In our laboratory, tissue culture techniques have been used successfully for propagation of several medicinally important plant species (Tsay [1999]; Nalawade et al. [2003]; Mulabagal and Tsay [2004]; Tsay and

Agrawal [2005]; Chen et al. [2006] and Chang et al. [2007]). Plants propagated by tissue culture have been reported to show less variation in the content of secondary metabolites than their cultivated or wild counterparts (Yamada et al. [1991]).

G. tenuifolia Cassini belonging to the family Asteraceae originates from Penghu Islands, Taiwan (Li[1978]). The perennial herb has been used to make traditional healthy food and herbal tea on the island for a long time. It is a traditional anti-pyretic and hepatoprotective used in the Chinese system of medicine (Anonymous [1999]). Plant decoction has been used for treatment of lung diseases, chronic nephritis, edema and prevention of sunstroke (Xu [1972]). The main active compounds of G. tenuifolia are luteolin, luteolin-7-glucoside and oleanolic acid (Anonymous [1999]). Several studies on G. tenuifolia have demonstrated its pharmacological values of antioxidation (Wu et al. [2005a]; Yang et al. [2006]), anti-inflammation (Wu et al. [2004]; Hsu et al. [2005]); immunomodulation (Ha et al. [2006]) and cytotoxicity on several human cancer cell lines (Hsu et al. [2005]). The ethanol extracts of G. tenuifolia showed strong ROS scavenging (antioxidant) activity in both cell free and cell based systems (Wu et al. [2005b]). Also, glossogin, an effective component of G. tenuifolia has been found to be a promising chemotherapeutic agent against lung cancer (Hsu et al. [2008]). It was demonstrated that the bioactive fraction 'Fr. C' of G. tenuifolia significantly inhibited the proliferation of A549 lung cancer cells (Hsu et al. [2011]). G. tenuifolia extract reportedly reduces the synthesis of the inflammatory mediator in activated murine macrophages RAW264.7 via an NF- B-dependent pathway (Wu et al. [2004]). G. tenuifoliaextract has also been found to inhibit the synthesis of a pro-inflammatory mediator in activated murine peritoneal macrophages, partially via NF-jB-dependent pathways (Ha et al. [2006]). In another study, G. tenuifolia extract suppressed the survival of mature osteoclasts by inhibiting osteoclast survival signaling pathways including NF-B, JNK, p38 and Akt, indicating its potential in the osteoclast-related diseases such as osteoporosis (Wang et al. [2014]).

Flavor of 'Xiang-Ru tea' (made from G. tenuifolia) is very popular in Taiwan. The Kaohsiung District Agricultural Improvement Station has developed canned 'Xiang-Ru tea' and 'Xiang-Ru jelly', and Penghu County farmers have come out with 'Xiang-Ru tea' bags. Additionally, other G. tenuifolia products and technologies are being developed, indicating an immense economic potential of the herb. However, the supply of natural plant material is seasonal and restricted to only a few months. Due to these constraints, production of *G. tenuifolia* plants having higher levels of active compounds and sustainability of its supply throughout the year are important issues. The development of a micropropagation protocol of *G. tenuifolia* offers the potential to alleviate these problems.

The purpose of this study was to develop a protocol for *in vitro* propagation of *G. tenuifolia*. The study also aimed to estimate the quantities of luteolin and oleanolic acid compounds in tissue culture raised plants; compare them with the quantities found in wild types and commercial crude drug using High-performance liquid chromatography (HPLC). The method developed in the present study can be applied for the mass production of true-to-type *G. tenuifolia* plants of superior genotypes at a commercial scale.

METHODS

Establishment of Aseptic Cultures

Seeds of *G. tenuifolia* were collected from Penghu Islands, Taiwan. Wild type plant materials were collected from Chimei and Wangan islands, Taiwan. Samples of commercial drug of *G. tenuifolia* were obtained from the authorized source. Seeds were surface disinfected by washing several times with sterile distilled water, followed by dipping in 70% (v/v) ethanol for 10s, then immersing in a solution of 1% (v/v) sodium hypochlorite containing 1 drop of Tween-20 for 5 min. Final washing consisted of 3 rinses of 5 min each in sterile distilled water. Thereafter, to raise *in vitro* seedlings, the disinfected seeds were cultured on ½X (half), and 1X (full) strength of macro, micro nutrients and vitamins of Murshige and Skoog's (Murashige and Skoog [1962]) medium, hereinafter referred as MS basal medium. The pH of the culture media was adjusted to 5.7 ± 0.1 before autoclaving. All media were gelled with 0.9% Difco Bacto-agar (Difco Laboratories, Detroit, MI, USA).

Multiple Shoots Induction

For induction of multiple shoots in *G. tenuifolia* , initial experiments were performed to find out a suitable culture medium, hence four different basal media, i.e. MS (Murashige and Skoog [1962]); WPM - Woody plant medium (Lloyd and McCown [1981]); B5 (Gamborg et al. [1968]); and N6 (Chu et al. [1975]); different concentrations of plant growth regulators (PGRs) e.g. BA (6-benzyladenine), Kin (Kinetin) and NAA (-naphthaleneacetic acid); and different concentrations of sucrose were tested in a randomized design. These trials gave us some idea of the growth regulators and the best basal medium needed to induce shoot induction in *G. tenuifolia* . Thereafter, shoot tips (0.8 to 1 cm long) were excised from 5 weeks old *in vitro* raised seedlings and were cultured on MS basal medium supplemented with 0.1, 0.5, 1.0 mg/L of 6-benzyladenine (BA) or 0.1, 0.5, 1.0 mg/L of kinetin and a fixed concentration (0.1 mg/L) of -naphthaleneacetic acid (NAA), 3% sucrose and 0.9% agar. The pH of all media was adjusted to 5.7 ± 0.1. Glass bottle (650 ml capacity), each having 100 ml of medium was used as culture vessel. After inoculation, the cultures were incubated in a growth room at $25 \pm °C$, with a light and dark period of 16/8 hr and a light intensity of 34 $\mu mol/m^2$ s. For fresh weight determination, the shoot cultures were gently pressed on sterile filter papers to remove excess water and weighed. The developing shoot clusters were sub-cultured onto the same medium composition every four weeks for further shoot proliferation and elongation of shoots.

Container Closure

This experiment was carried out to test the influence of the type of container closure optimum for *in vitro* shoot cultures of *G. tenuifolia*. Glass bottles used as culture vessels were closed (sealed) with 4 dispensable papers (DP4) or with 2 layers of aluminum foil (AF2). In case of DP4, culture containers were additionally closed with a layer of parafilm which was removed after 2 weeks to facilitate a better gas exchange.

Rooting of *In Vitro* Shoots

For induction of rooting, *in vitro* shoots were cultured in ½X MS basal medium supplemented with Indole-3-acetic acid (IAA) at 0.5, 1, 3 and 5 mg/L concentrations or Idole-3-butyric acid (IBA) at 0.05, 0.1, 0.5 and 1 mg/L concentrations. Sucrose (3%) and agar (0.9%) was added to all the media. Observations were recorded after 35 days of culture.

Hardening and Survival of Tissue Culture Plants

Rooted plantlets were removed from the culture vessels, rinsed with water to remove the medium, and then transferred to plastic pots (9 cm diameter) containing a mixture of peat soil: perlite: vermiculite (1:1:1 v/v) in the greenhouse. The plants were watered once a day. Initially, a higher humidity was maintained by keeping the pots in a tray having water. Each pot was covered with a thin transparent polyethylene bag (sachet). After one week, one top corner of the sachet was cut. During the third week, a similar cut was made on the other side of the sachet. This sachet was completely removed in the fourth week. During the fifth week, pots were taken out of the tray. The data on survival of plants was recorded after five weeks of transfer to the greenhouse.

HPLC Analysis of Tissue Culture Plants and Commercial Crude Drug

Reagents, Materials and Conditions

HPLC-grade methanol was purchased from Merck (Germany). Pump (L-2130), auto injector (L-2200) and diode array detector system (L-2450) used were from Hitachi. Symmetry Water Column C_{18} (5 µm, 4.6 × 250 mm) and Milli Q water (Millipore, Milford, MA, USA) were used for all the analysis carried out at room temperature. The mobile phase for luteolin was a gradient eluting with acetonitrile/water (0.5% acetic acid) (from 20:80 to 100:0, by v/v) at 1 ml/min over 47 min. The eluent was monitored at 245 nm. While, mobile phase for oleanolic acid was a mixture of reagent acetonitrile and water (05% acetic acid) (90:10) at 1 ml/min over 15 min. The eluent was monitored at 210 nm.

Preparation of HPLC Standard and Samples

Luteolin (ChromaDex) and oleanolic acid (Fluka Analytical) standard samples were purchased from Sigma-Aldrich Co. LLC. Standard solutions were prepared by dissolving 5 mg of each in 5 ml of Ethanol. Dissolved solutions (1.0 mg ml^{-1}) were filtered through a 0.22 µm filter (Millipore, USA) and further diluted to the concentration of

- For Luteolin: 1.0, 2.0, 4.0, 8.0 and 16 mg l^{-1}.
- For Oleanolic acid: 5.0, 10.0, 20.0, 40.0, 80.0 and 160.0 mg l^{-1}.

Calibration curves for standards were established and high linearity (r2 > 0.999) was obtained for each calibration curve. Standard solution (10 µl) was used for HPLC injections. Calibration graphs were plotted based on linear regression analyses of the peak areas in response to concentrations of standards injected. The repeatability of the migration time and peak areas of Luteolin and oleanolic acid in the experiment were determined under the optimum conditions. Samples of *in vitro* shoots for the HPLC analysis were collected from the culture vessels and their fresh weights were recorded. The samples were then freeze-dried for 24 h and their dry weights were determined. Fraction (100 mg) of each dried sample was crushed into fine powder and dissolved in 10 ml of ethanol. It was ultra-sonicated for 10 min and the supernatant was collected after centrifugation (5000 rpm, 10 min). This process was repeated three times for each sample. After filtration, the combined ethanol extracts were evaporated to dryness with the help of a rotary evaporator. The residue was dissolved in 10 ml ethanol and filtered through a 0.22 µm (Millipore, USA) membrane before analysis.

Statistical Analysis

Each treatment consisted of thirty replicates and each experiment was repeated three times. Data were statistically analyzed for least significant difference (LSD) using SAS 8.2 statistical software (SAS Institute Inc [2001]).

RESULTS AND DISCUSSION

Induction of Multiple Shoots

Surface-sterilized seeds of *G. tenuifolia* inoculated on different basal media commenced gemination after 3 days of inoculation and a maximum 50% of seed germination at day 15 was recorded. The induction of multiple shoots in explants varied with cytokinin type and concentration (Table 1) (Figure 1). The response in medium supplemented with BA (1.0 mg/L) + NAA (0.1 mg/L) was higher compared to the medium with kinetin + NAA, or devoid of growth regulators. The maximum average number of shoots (7.4 ± 0.6) in 100% explants was obtained on ½X MS medium supplemented with BA (1.0 mg/L) + NAA (0.1 mg/L) and 3% sucrose after 35 days of inoculation. Also, this treatment resulted in the maximum fresh weight (894.9 ± 78.4) per explant. Multiple shoots developed directly from the lateral bud meristems. Kinetin and medium without plant growth regulators though supported elongation of shoots, but responses were lower compared to BA which induced a higher number of multiple shoots at all three concentrations (0.1, 0.5 and 1.0 mg/L) (Table 1).

Table 1: Influence of BA and Kin on induction of multiple shoots in seedling-derived shoot tip explants of *G. Tenuifolia*

Cytokinin* (mg/L)	NAA (mg/L)	No. of explants cultured	Shoot length (mm)**	Explants induced multiple Shoots (%) **	Average No. of shoots / explant**	Explants induced callus (%) **	Fresh weight/ shoot (mg)**
0	0	30	26.7±2.1 cd	53.3±9.1 bc	1.8±0.3 c	0.0±0.0 d	128.0±16.7 d
0	0.1	30	40.8±2.2 a	50.0±9.1 bc	1.8±0.2 c	56.7±9.0 b	379.5±33.4 c
BA 0.1	0.1	30	24.4±1.3 d	93.3±4.6 a	4.1±0.4 b	33.3±8.6 c	364.1±43.6 c
BA 0.5	0.1	30	24.2±1.4 d	100.0±0.0 a	7.3±0.6 a	13.3±6.2 d	666.2±68.4 b
BA 1.0	0.1	30	26.2±1.7cd	100.0±0.0 a	7.4±0.6 a	95.0±4.9 a	894.9±78.4 a
Kin 0.1	0.1	30	40.1±3.5 a	43.3±9.0 c	1.8±0.2 c	50.0±9.1 bc	319.8±44.2 c
Kin 0.5	0.1	30	32.8±2.4 bc	70.0±8.4 b	2.5±0.3 c	80.0±7.3 a	306.7±32.6 c
Kin 1.0	0.1	30	34.6±2.1 ab	56.7±9.0 bc	2.3±0.3 c	100.0±0.0 a	253.0±26.0cd

*Basal medium: ½X MS basal medium supplemented with 3% sucrose and 0.9% Difco Bacto-agar. Observations were recorded after 35 days of culture.

**Means followed by the same letter are not significantly different at 5% level by LSD (least significant difference) test.

Chen *et al.*

Chen *et al. Botanical Studies* 2014 **55**:45, doi: 10.1186/s40529-014-0045-7

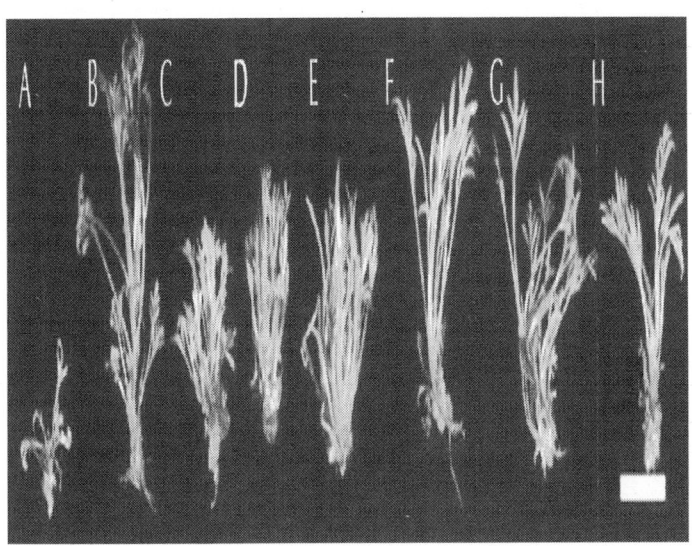

Figure 1: Influence of BA and Kinetin on induction of multiple shoots in seedling-derived shoot tip of *G. tenuifolia*. (A) ½X MS basal medium without PGRs; (B) 0.1 mg/L NAA; (C) 0.1 mg/L BA and 0.1 mg/L NAA; (D) 0.5 mg/L BA and 0.1 mg/L NAA; (E) 1 mg/L BA and 0.1 mg/L NAA; (F) 0.1 mg/L Kin and 0.1 mg/L NAA; (G) 0.5 mg/L Kin and 0.1 mg/L NAA; (H) 1 mg/L Kin and 0.1 mg/L NAA. (Bar = 1 cm).

Similar to the results in the present study, the differential effect of various concentrations of BAP on the induction of multiple shoots has earlier been reported for *Gossypium* (Agrawal et al.[1997]), *Salix* (Agrawal and Gebhardt [1994]), *Pisum* (Jackson and Hobbs [1990]) *Phaseolus* (McClean and Grafton [1989]) and *Glycine* (Cheng et al. [1980]). Akin to the present study, BAP was the most effective cytokinin in all these reports, indicating a particular cytokinin preference for multiple shoot induction in certain tissues.

Container Closure

Our experiment to find out the type of container closure optimum for *in vitro* shoot cultures of *G. tenuifolia* showed that the maximum average number of multiple shoots (9.7/ explant) and average fresh weight of each shoot (472.1 mg) was obtained when each culture container was closed with 2 layers of aluminum foil (AF). The container closure with 4 dispensable papers (DP4) resulted in a lower average number of shoots to 7.1/explant and fresh weight of 326.2 mg (Table 2) (Figure 2).

Table 2: Influence of container closure type on induction of multiple shoots in *G. Tenuifolia*

Ventilation closure*	Culture medium	Explants induced multiple shoots (%)***	**Average No. of shoots / explant *****	**Fresh weight / shoot (mg) *****
AF2	½ X MS basal	61.4 ± 6.2 b	2.3 ± 0.2 c	128.7 ± 6.8 c
AF2	NAA (0.1) + BA (0.1)	100.0 ± 0.0 a	9.7 ± 0.4 a	472.1 ± 15.2 a
DP4 **	½ X MS basal	68.6 ± 2.5 b	2.5 ± 0.1 c	73.5 ± 2.8 d
DP4	NAA (0.1) + BA (0.1)	100.0 ± 0.0 a	7.1 ± 0.5 b	326.2 ± 12.2 b

*Basal medium: ½ X MS basal salts supplemented with 3% sucrose and 0.9% Difco Bacto-agar. Concentrations of PGRs in the parentheses represent mg/L values. Observations were recorded after 35 days of culture.

** AF2 = Culture container closed with 2 layers of Aluminum foil;

DP4 = Culture container sealed with 4 dispense papers. Culture containers were initially sealed with 4 dispense paper and parafilm layers. After 2 weeks, parafilm layer was removed to facilitate ventilation.

***Means followed by the same letter are not significantly different at 5% level by LSD (least significant difference) test.

Chen *et al.*

Chen *et al. Botanical Studies* 2014 **55**:45, doi: 10.1186/s40529-014-0045-7

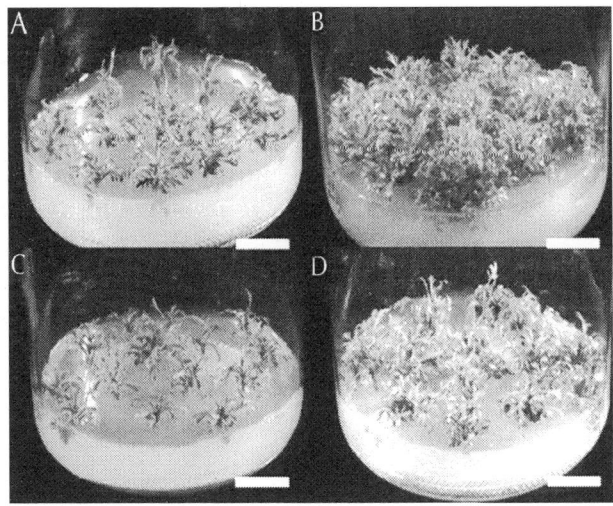

Figure 2: Influence of container closure type on induction of multiple shoots in *G. tenuifolia*. (A) Sealed with 2 layers of AF; ½X MS basal (B) Sealed with 2 layers of AF; BA 0.1 mg/L + 0.1 mg/L (C) Sealed with 4 layers of DPs; ½X MS basal; (D) Sealed with 4 layers of DPs; BA 0.5 mg/L + NAA 0.1 mg/L. (Bar = 2 cm).

To maintain the sterility of cultures, it is essential to close culture containers with some closure. Different types of container closures are commonly used. Some closures cause restriction of gaseous exchange between the container atmosphere and the outside environment (Buddendorf-Joosten and Woltering [1994]), which can result in poor aeration and hyperhydric condition of cultures. It has been reported that the type of closure affects gaseous exchange, availability of water, micronutrients, and balance of hormones in the culture container (Kataeva et al. [1991]; Lai et al. [2005]; Chen et al. [2006]; Tsay et al. [2006]). Also, growth rate and other physiological and morphological characteristics of plants developed under *in vitro* conditions can be influenced by the physical and chemical micro-environments of culture containers (Walker et al. [1988]). Different species show different requirement with respect to container closures. Hence, it is important to optimize a closure type in a micropropagation protocol of a particular plant species.

Rooting of *in vitro* Shoots

Between the two auxins tested, the response of IAA was found better compared to IBA for induction of rooting in *in vitro* shoots of *G. tenuifolia*. Half strength (½X) MS basal medium supplemented with 3.0 mg/L IAA induced an average number of 6.6 roots/shoot in 100% shoots (Table 3). The response of IBA in the medium was also comparable (6.3 roots/shoot), however, with IAA, the percentage of rooting was lower (83.3). Higher concentrations of both the auxins (>3.0 mg/L IAA and > 0.1 mg/L IBA) induced more average number of roots per shoot, however, shoots induced callus at the base which affected the survival rates of the rooted shoots in the greenhouse (Table 3). The promotory effect of a lower salt concentration of MS on *in vitro* rooting of shoots has been reported for *Gossypium* (Agrawal et al. [1997]), *Philodendron* spp. (Maene and Debergh [1985]). Different plant species respond differently to auxins for the induction of rooting. Some plant species, even do not require any auxin supplemental in the medium for rooting (Agrawal and Gebhardt [1994]), hence it is desirable to optimize the type and concentration of an auxin in a micropropagation protocol of a particular plant species.

Table 3: Influence of IAA and IBA concentrations on *in vitro* rooting of shoots of *G. Tenuifolia*

Auxin (mg/L) *	Concentration (mg/L)	No. of shoots rooted (%) **	Average No. of roots **	Callus formation	Plants survival in greenhouse (%)
IAA	0.5	53.3 ± 9.1 b	2.4 ± 0.4 cd	-	93.3
-	1.0	63.3 ± 8.8 b	3.3 ± 0.5 c	-	100.0
-	3.0	100.0 ± 0.0 a	6.6 ± 0.7 b	+	100.0
	5.0	93.3 ± 4.6 ab	9.5 ± 0.9 a	++	96.7
IBA	0.05	3.3 ± 3.4 d	0.1 ± 0.1 e	-	90.0
	0.1	16.7 ± 6.8 c	0.9 ± 0.3 de	-	96.7
	0.5	83.3 ± 6.8 ab	6.3 ± 0.7 b	+	96.7
	1.0	96.7 ± 3.3 ab	8.6 ± 0.7 a	+++	66.7

*Basal medium: **½X MS basal medium supplemented with 3% sucrose and 0.9%** Difco Bacto-agar. Observations were recorded after 35 days of culture.

**Means followed by the same letter are not significantly different at 5% level by LSD (least significant difference) test.

Chen *et al.*

Chen *et al. Botanical Studies* 2014 **55**:45, doi: 10.1186/s40529-014-0045-7

Hardening and Survival of Tissue Culture Plants

Hardening and 100% survival of tissue culture plants was achieved on the peat soil: perlite: vermiculite (1:1:1 v/v) mix in plastic pots kept in the greenhouse. Covering of plants with transparent sachets raised humidity levels, crucial for the survival of tissue culture plants of *G. tenuifolia*. Normal flowering and seed formation (Figure 3A, B) was observed in all the tissue culture raised plants after 3½ months of transfer to pots.

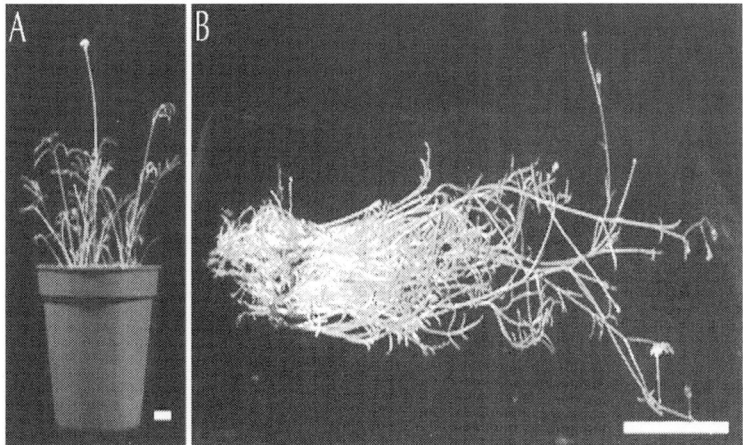

Figure 3: (A) Tissue culture plants of *G. Tenuifolia* successfully acclimatized, (B) Tissue culture plants in greenhouse (3½ month old). Bar A = 1.3 cm;Bar B = 6 cm).

HPLC Analysis of Secondary Metabolites

HPLC analysis revealed the varying quantities of oleanolic acid and luteolin in *in vitro* shoots, tissue culture plants in the greenhouse, wild type and commercial crude drug materials (Table 4). The oleanolic acid and luteolin contents were found to be significantly higher (16.89 mg/g and 0.84 mg/g, respectively) in 3½-month old tissue culture raised plants in greenhouse compared to commercially available crude drug (6.51 mg/g, 0.13 mg/g, respectively). In comparison with the tissue culture plants in the greenhouse, the next best values of oleanolic acid and luteolin contents were recorded in wild type plants from Chimei and Wangan islands. There was only a marginal difference between the two wild type plant materials showing comparable quantities of both the active compounds. These compounds were present mostly in the aboveground parts of the plant. Only underground dried herbs of commercial crude drug showed oleanolic acid content, indicating probable mixing of aboveground parts with the underground parts or a wrong identification of the commercial crude drug. *In vitro* shoots growing under culture conditions on ½ strength MS basal medium devoid of any growth regulators also showed the presence of active compounds (3.29 mg/g of oleanolic acid and 0.47 mg/g of luteolin) indicating onset of biosynthetic pathways for production of these compounds at the culture stage itself. The lower quantities of the compounds in *in vitro* shoots in comparison to the tissue culture raised greenhouse plants may be due to differences in the maturity of the plant materials. Media supplemented with PGRs did not have significant influence on the contents of oleanolic acid and luteolin in *in vitro* shoots (Data not shown).

Table 4: HPLC analysis for luteolin and oleanolic acid contents in *in vitro* raised, wild and commercially available plant materials of *Glossogyne tenuifolia*

Plant Samples	Source / Treatment	Oleanolic acid (mg/g of dw)*	Luteolin (mg/g of dw)*
Commercial crude drug	Dried herbs(Aboveground)	6.51	0.13
	Dried herbs(Underground)	1.07	none

Wild type	Chimei Island (Aboveground)	13.78	0.82
	Chimei Island (Underground)	none	none
	Wangan Island (aboveground)	14.58	0.72
	Wangan Island (Underground)	none	none
In vitro **shoots**	**½X MS basal medium** (−PGRs)	3.29	0.47
Tissue culture plants (3 month old) in greenhouse	Aboveground parts	16.89	0.84
	Underground parts	none	none

dw: Freeze-dried weight

Chen *et al.*

Chen *et al. Botanical Studies* 2014 55:45, doi: 10.1186/s40529-014-0045-7

Similar to the present study, significantly higher amounts of emodin and physcion contents were observed in *in vitro* propagated shoots and tissue culturplants of *Polygonum multiflorum* compared to the marketed crude drug (Lin et al. [2003]). Fairly high amounts of gentiopicroside and swertiamarin compounds were recorded in the aerial and underground parts of *Gentiana davidii*var. *formosana* compared to commercially available crude drug (Chueh et al. [2001]). Yet in another study on *Saussurea involucrata* in our laboratory, the highest Syringetin and Rutin contents were recorded in petioles of two months old *in vitro* plants compared to the commercially available crude drug (unpublished results). The possible reasons for the enhanced levels of active compounds in tissue culture plants could be, that the plants collected for the crude drugs are grown under natural conditions and the contents of the active compounds may vary depending on the growth conditions, place and time of plant collections. Whereas, the tissue culture plants are grown under controlled growth and environmental conditions receiving an optimum supply of nutrients and other favorable growth conditions. Based on the results of the present study, it is evident that under defined culture conditions, it is possible to produce plants with higher contents of oleanolic acid and luteolin contents in a shorter time span and throughout the year.

CONCLUSIONS

In the present study, we have developed an *in vitro* propagation protocol for *G. tenuifolia*, an important medicinal plant native to Taiwan. HPLC analysis of tissue culture raised plants grown in the greenhouse have shown significantly higher levels of the active compounds compared to wild types and commercial crude drug, demonstrating the usefulness of the tissue culture technology. The results obtained in the present study have enormous significance, since so far there is no published report on micropropagation of this medicinally important plant species.

AUTHORS' CONTRIBUTION

CCC carried out the experimental work, CHC provided funding and designed the experiments, CLK identified the wild plant materials and provided other significant inputs during the study, DCA prepared the manuscript, CRW helped in analytical work, and HST provided laboratory facilities and important inputs in the manuscript preparation. All authors read and approved the final manuscript.

ACKNOWLEDGEMENTS

Research grant (NSC 100-2313-B-324-002) by the National Science Council, Taiwan is gratefully acknowledged.

REFERENCES

1. Agrawal DC, Gebhardt K (1994) Rapid micropropagation of hybrid willow (*Salix*) established by ovary culture. J Plant Physiol 143(6):763-765

2. Agrawal DC, Banerjee AK, Kolala RR, Dhage AB, Nalawade SM, Kulkarni AV, Hazra S, Krishnamurthy KV (1997) In vitro induction of multiple shoots and plant regeneration in cotton (*Gossypium hirsutum* L.). Plant Cell Rep 16:647-652 [10.1007/BF01275508] doi: 10.1007/BF01275508

3.   (1999) Zhong Hua Ben Cao (China Herbal). Shanghai Science and Technology Press, Shanghai.

4.   Buddendorf-Joosten JMC, Woltering EJ (1994) Components of the gaseous environment and their effects on plant growth and development *in vitro*. Plant Growth Regul 15:1-16

5.   Cheng TY, Saka H, Voqui-Dinh TH (1980) Plant regeneration from soybean cotyledonary node segments in culture. Plant Sci Lett 19:91-99

6.   Chang HC, Agrawal DC, Kuo CL, Wen JL, Chen CC, Tsay HS (2007) *In vitro* culture of*Drynaria fortunei*, a fern species source of Chinese medicine "Gu-Sui-Bu". In Vitro Cell Dev Biol-Plant 43:133-139   [10.1007/s11627-007-9037-6]   doi: 10.1007/s11627-007-9037-6

7.   Chen UC, Hsia CN, Agrawal DC, Tsay HS (2006) Influence of ventilation closures on plant growth parameters, acclimation and anatomy of leaf surface in *Scrophularia yoshimurae*Yamazaki - a medicinal plant native to Taiwan. Bot Stud 47:259-266

8.   Chu CC, Wang CC, Sun CS, Hsu C, Yin KC (1975) Establishment of an efficient medium for anther culture of rice through comparative experiments on the nitrogen sources. Sci Sin 18:659-668

9.   Chueh FS, Chen CC, Sagare AP, Tsay HS (2001) Quantitative determination of secoiridoid glucosides in in vitro propagated plants of *Gentiana davidii* var. *formosana* by high performance liquid chromatography. Planta Med 67:70-73

10.   Cotell N, Bernier JL, Catteau JP, Pommery J, Wallet JC, Gaydou EM (1996) Antioxidant properties of hydroxy-flavones. Free Radic Biol Med 20:35-43

11.   Craig WJ (1999) Health-promoting properties of common herbs. Am J Clin Nutr 70:491S-499S

12.   Gamborg OL, Miller RA, Ojima K (1968) Nutrient requirements of suspension cultures of soybean root cells. Exp Cell Res 50:151-158   [10.1016/0014-4827(68)90403-5]   doi: 10.1016/0014-4827(68)90403-5.

13.   Ha CL, Weng CY, Wang L, Lian TW, Wu MJ (2006) Immunomodulatory effect of *Glossogyne tenuifolia* in murine peritoneal macrophages and splenocytes. J Ethnopharmacol 107:116-125

14. Hoareau L, DaSilva E (1999) Medicinal plants: a re-emerging health aid. EJB Electronic J Biotech 2:56-70

15. Hsu HF, Houng JY, Chang CL, Wu CC, Chang FR, Wu YC (2005) Antioxidant activity, cytotoxicity, and DNA information of *Glossogyne tenuifolia*. J Agric Food Chem 53:6117-6125

16. Hsu HF, Houng JY, Kuo CF, Tsao N, Wu YC (2008) Glossogin, a novel phenylpropanoid from*Glossogyne tenuifolia*, induced apoptosis in A549 lung cancer cells. Food Chem Toxicol 46:3785-3791

17. Hsu HF, Wu YC, Chang CC, Houng JY (2011) Apoptotic effects of bioactive fraction isolated from *Glossogyne tenuifolia* on A549 human lung cancer cells. J Taiwan Inst Chem Eng 42:556-562

18. Jackson JA, Hobbs SLA (1990) Rapid multiple shoot production from cotyledonary node explants of Pea (*Pisum sativum*). In Vitro Cell Dev Biol 26:835-838

19. Kataeva NV, Alexandrova IG, Butenko RG, Dragavtceva EV (1991) Effect of applied and internal hormones on vitrification and apical necrosis of different plants cultured *in vitro*. Plant Cell Tiss Organ Cult 14:31-40

20. Lai CC, Lin HM, Nalawade SM, Fang W, Tsay HS (2005) Hyperhydricity in shoot cultures of*Scrophularia yoshimurae* can be effectively reduced by ventilation of culture vessels. J Plant Physiol 162(3):355-361

21. Li HL (1978) Glossogyne Cass. Flora of Taiwan. Epoch Publishing Co, Taipei, Taiwan.

22. Lin LC, Nalawade SM, Mulabagal V, Yeh MS, Tsay HS (2003) Micropropagation of *Polygonum multiflorum* THUNB and quantitative analysis of the anthraquinones emodin and physcion formed in *in vitro* propagated shoots and plants. Biol Pharm Bull 26(10):1467-1471

23. Lloyd G, McCown B (1981) Commercially feasible micropropagation of mountain laurel,*Kalmia latifolia*, by shoot tip culture. Intl Plant Prop Soc Proc 30:421-427

24. Maene L, Debergh P (1985) Liquid medium addition to established tissue cultures to improve elongation and rooting in vitro. Plant Cell Tissue Organ Cult 5:23-33

25. McClean P, Grafton KF (1989) Regeneration of dry bean (*Phaseolus vulgaris* L.) via organogenesis. Plant Sci 60:117-122

26. Mulabagal V, Tsay HS (2004) Plant cell cultures-An alternative and efficient source for the production of biologically important secondary metabolites. Int J Appl Sci Eng 2(1):29-48

27. Murashige T, Skoog F (1962) A revised medium for rapid growth and bioassays with tobacco tissue cultures. Physiol Plant 15:473-497 [10.1111/J.1399-3054.1962.TB08052.X] doi: 10.1111/J.1399-3054.1962.TB08052.X

28. Nalawade SM, Sagare AP, Lee CY, Kuo CL, Tsay HS (2003) Studies on tissue culture of Chinese medicinal plant resources in Taiwan and their sustainable utilization. Bot Bull Acad Sinica 44:79-98

29. (2001) SAS/STAT User's Guide. Version 8.2, vol 2. SAS Institute, USA.

30. Tsay HS, Agrawal DC (2005) Tissue culture technology of Chinese medicinal plant resources in Taiwan and their sustainable utilization. Int J Appl Sci Eng 3:215-223

31. Tsay HS (1999) Tissue culture technology of medicinal herbs and its application in Taiwan. In: Chou CH, Waller GR, Reinhardt C (eds) Biodiversity and allelopathy: from organisms to ecosystems in the Pacific, Academia Sinica, Taipei. pp 137-144

32. Tsay HS, Lee CY, Agrawal DC, Basker S (2006) Influence of ventilation closure, gelling agent and explant type on shoot bud proliferation and hyperhydricity in *Scrophularia yoshimurae* – A medicinal plant. In Vitro Cell Dev Biol-Plant 42:445-449

33. Vieira RF, Skorupa LA (1993) Brazilian medicinal plants gene bank. Acta Hort 330:51-58

34. Walker PN, Heuser CW, Heinemann PH (1988) Micropropagation: studies of gaseous environments. Acta Hort 230:145-151

35. Wang SW, Kuoa HC, Hsu HF, Tu YK, Chenge TT, Houng JY (2014) Inhibitory activity on RANKL-mediated osteoclastogenesis of *Glossogyne tenuifolia* extract. J Funct Food 6:215-223

36. Wu MJ, Wang L, Ding HY, Weng CY, Yen JH (2004) *Glossogyne tenuifolia* acts to inhibit inflammatory mediator production in a macrophage cell line by down-regulating LPS-induced NF- B. J Biomed Sci 11(2):186-199

37. Wu MJ, Huang CL, Lian TW, Kou MC, Wang L (2005) Antioxidant activity of *Glossogyne tenuifolia*. J Agric Food Chem 53:6305-6312

38. Wu MJ, Weng CY, Ding HY, Wu PJ (2005) Anti-inflammatory and antiviral effects of *Glossogyne tenuifolia*. Life Sci 76:1135-1146

39. Yang JH, Tsai SY, Han CM, Shih CC, Mau JL (2006) Antioxidant properties of *Glossogyne tenuifolia*. Am J Chin Med 34:707-720

40. Xu HY (1972) Illustrations of Chinese herbs in Taiwan. Department of Health, Executive Yuan, Taiwan.

41. Yamada Y, Shoyama Y, Nishioka I, Kohda H, Namera A, Okamoto T (1991) Clonal micropropagation of *Gentiana scabra* Bunge var. *buergeri* Maxim. and examination of the homogeneity concerning the gentiopicroside content. Chem Pharm Bull 39:204-206

Citations

CHAPTER 1

Loiy Elsir A Hassan, Mohamed B Khadeer Ahamed, Aman S Abdul Majid, Hussein M Baharetha, Nahdzatul S Muslim, Zeyad D Nassar, and Amin MS Abdul Majid, Correlation of Antiangiogenic, Antioxidant and Cytotoxic Activities of some Sudanese Medicinal Plants with Phenolic and Flavonoid Contents, doi:10.1186/1472-6882-14-406.

CHAPTER 2

Tavasalkar SU, Mishra HN, Madhavan S (2012) Evaluation of Antioxidant Efficacy of Natural Plant Extracts against Synthetic Antioxidants in Sunflower Oil. 1:504, doi:10.4172/scientificreports.504.

CHAPTER 3

Mehar Darukhshan Kalim, Dipto Bhattacharyya, Anindita Banerjee, and Sharmila Chattopadhyay, Oxidative DNA damage preventive activity and antioxidant potential of plants used in Unani system of medicine, doi:10.1186/1472-6882-10-77.

CHAPTER 4

D. Sreeramulu, C. V. K. Reddy, Anitha Chauhan, N. Balakrishna, and M. Raghunath, "Natural Antioxidant Activity of Commonly Consumed Plant Foods in India: Effect of Domestic Processing," Oxidative Medicine and Cellular Longevity, vol. 2013, Article ID 369479, 12 pages, 2013, doi:10.1155/2013/369479.

CHAPTER 5

Md. Kamal Uddin, Abdul Shukor Juraimi, Md Sabir Hossain, Most. Altaf UN Nahar, Md. Eaqub Ali, and M. M. Rahman, "Purslane Weed (Portulaca oleracea): A Prospective Plant Source of Nutrition, Omega-3 Fatty Acid, and Antioxidant Attributes," The Scientific World Journal, vol. 2014, Article ID 951019, 6 pages, 2014. doi:10.1155/2014/951019.

CHAPTER 6

Yasmin Hilmi, Muna F Abushama, Haidar Abdalgadir, Asaad Khalid, and Hassan Khalid, A Study of Antioxidant Activity, Enzymatic Inhibition and In Vitro Toxicity of Selected Traditional Sudanese Plants With Anti-Diabetic Potential, doi:10.1186/1472-6882-14-149.

CHAPTER 7

Kuan-Hung Lin, Yan-Yin Yang, Chi-Ming Yang, Meng-Yuan Huang, Hsiao-Feng Lo, Kuang-Chuan Liu, Hwei-Shen Lin, and Pi-Yu Chao, Antioxidant Activity of Herbaceous Plant Extracts Protect Against Hydrogen Peroxide-Induced DNA Damage in Human Lymphocytes, doi:10.1186/1756-0500-6-490.

CHAPTER 8

Tsung-Hsien Tsai, Ching-Jang Huang, Wen-Huey Wu, Wen-Cheng Huang, Jong-Ho Chyuan, and Po-Jung Tsai, Antioxidant, cell-protective, and anti-melanogenic activities of leaf extracts from wild bitter melon (Momordica charantiaLinn. var.abbreviata Ser.) cultivars, doi: 10.1186/s40529-014-0078-y.

CHAPTER 9

Chia-Chen Chen, Hung-Chi Chang, Chao-Lin Kuo, Dinesh Chandra Agrawal, Chi-Rei Wu, and Hsin-Sheng Tsay, In Vitro Propagation and Analysis of Secondary Metabolites in Glossogyne Tenuifolia (Hsiang-Ju) - A Medicinal Plant Native to Taiwan, doi:10.1186/s40529-014-0045-7.

Index